Constraining Designs for Synthesis
and Timing Analysis

WITHDRAWN
UTSA LIBRARIES

Sridhar Gangadharan · Sanjay Churiwala

Constraining Designs for Synthesis and Timing Analysis

A Practical Guide to Synopsys Design Constraints (SDC)

with Chapter 17 contributed by
Frederic Revenu

Sridhar Gangadharan
Atrenta, Inc.
San Jose, CA, USA

Sanjay Churiwala
Xilinx
Hyderabad, India

ISBN 978-1-4614-3268-5   ISBN 978-1-4614-3269-2 (eBook)
DOI 10.1007/978-1-4614-3269-2
Springer New York Heidelberg Dordrecht London

Library of Congress Control Number: 2013932651

© Springer Science+Business Media New York 2013
This work is subject to copyright. All rights are reserved by the Publisher, whether the whole or part of the material is concerned, specifically the rights of translation, reprinting, reuse of illustrations, recitation, broadcasting, reproduction on microfilms or in any other physical way, and transmission or information storage and retrieval, electronic adaptation, computer software, or by similar or dissimilar methodology now known or hereafter developed. Exempted from this legal reservation are brief excerpts in connection with reviews or scholarly analysis or material supplied specifically for the purpose of being entered and executed on a computer system, for exclusive use by the purchaser of the work. Duplication of this publication or parts thereof is permitted only under the provisions of the Copyright Law of the Publisher's location, in its current version, and permission for use must always be obtained from Springer. Permissions for use may be obtained through RightsLink at the Copyright Clearance Center. Violations are liable to prosecution under the respective Copyright Law.
The use of general descriptive names, registered names, trademarks, service marks, etc. in this publication does not imply, even in the absence of a specific statement, that such names are exempt from the relevant protective laws and regulations and therefore free for general use.
While the advice and information in this book are believed to be true and accurate at the date of publication, neither the authors nor the editors nor the publisher can accept any legal responsibility for any errors or omissions that may be made. The publisher makes no warranty, express or implied, with respect to the material contained herein.

Printed on acid-free paper

Springer is part of Springer Science+Business Media (www.springer.com)

Foreword

It has been said that "timing is everything." While that is certainly true if you're in show business, the same holds true if you're designing a system-on-a-chip (SoC). SoCs are powering the hand-held consumer electronics revolution going on all around us. They make things like smart phones and tablets possible. Correct definition and management of timing constraints for an SoC are critical tasks. How well these tasks are done will impact the success of the chip project.

An SoC is typically a collection of many complex building blocks sourced from multiple suppliers. It is the designer's job to stitch all these blocks together and achieve the sometimes competing goals of power, performance, and cost for the chip. And all of this happens while the whole team is under tremendous schedule pressure. The fact that so many SoC devices work the first time is nothing short of a miracle. There are many challenges associated with SoC design and many significant technologies that help make them possible.

In the chapters that follow, Sridhar Gangadharan and Sanjay Churiwala take an in-depth look at timing constraints. The broad impact that timing constraints have on the success of an SoC design project is discussed. Many examples are presented for both ASIC and FPGA design paradigms. On the surface, defining timing constraints appears to be a straightforward process. It is, in fact, a complex process with many important nuances and interrelationships. Sridhar and Sanjay do an excellent job explaining the process with many relevant examples and detailed "how to" explanations.

As designs have grown in complexity, much effort has gone into initiatives focused on improving design efficiency and managing risk. What is not fully understood is the impact that timing constraints have on both. Poorly managed or incorrect constraints can have significant negative impact on design effort and can lead to a chip failure. The chances of this occurring are growing with every new technology node. I believe that timing constraints are coming upon us as a major area of design challenge, and I congratulate Sridhar and Sanjay for developing such a complete guide for this important topic. I hope you find it useful as well.

Dr. Ajoy Bose
Chairman, President and CEO, Atrenta Inc.
San Jose, CA, USA

Preface

Dear Friends,

In today's world of deep submicron, Timing has become a critical challenge for designers developing Application Specific Integrated Circuits (ASIC) or System on a chip (SoC). Design engineers spend many cycles iterating between different stages of the design flow to meet the timing requirements. Timing is not merely a response time of a chip, but an integral part of the chip functionality that ensures that it can communicate with other components on a system seamlessly. That begs the question, what is timing? How do you specify it?

This book serves as a hands-on guide to writing and understanding timing constraints in integrated circuit design. Readers will learn to write their constraints effectively and correctly, in order to achieve the desired performance of their IC or FPGA designs, including considerations around reuse of the constraints. Coverage includes key aspects of the design flow impacted by timing constraints, including synthesis, static timing analysis, and placement and routing. Concepts needed for specifying timing requirements are explained in detail and then applied to specific stages in the design flow, all within the context of Synopsys Design Constraints (SDC), the industry-leading format for specifying constraints.

We have often heard from many design engineers that there are several books explaining concepts like Synthesis and Static Timing Analysis which do cover timing constraints, but never in detail. This book is our attempt at explaining the concepts needed for specifying timing requirements based on many years of work in the areas of timing characterization, delay calculation, timing analysis, and constraints creation and verification.

Book Organization

Here's how the book is laid out:

Chapters 1, 2, and 3 introduce the subject of Timing Analysis – including its need in the context of design cycle. The descriptions in these chapters are vendor, language, and format-independent.

Chapter 4 provides an overview of the Tcl language, because SDC (Synopsys Design Constraints) acts as an extension to Tcl. The concept of SDC is also introduced in this chapter.

These first four chapters might be thought of as Introduction section.

Chapters 5 through 8 together form a section which talks about clocks, explaining how to apply clock-related constraints. These chapters explain various kinds of clocks and their relationships and how to specify those in SDC.

Chapters 9 and 10 explain how to apply constraints on the remaining (non-clock) ports. With this section, all the primary ports are covered.

Chapters 11, 12, and 13 explain the need for timing exceptions. These chapters then go on to explain how to specify the exceptions correctly in SDC.

Chapters 14 and 15 deal with much more specialized topics. These concepts are less about individual constraints. Rather they delve into how design teams manage the world of constraints as they move across the flow, from front-end to back-end, partitioning the complete design to blocks and when integrating individual blocks.

In Chap. 16, we explain some other commands of SDC, which might have an impact on Timing Analysis.

Some of the commands are still not covered in this book. However, with the fundamental understanding gained on Timing Analysis and SDC through these chapters, it should be possible for a user to easily comprehend any remaining commands, including any extensions that might come in future versions of SDC.

Most tools which support SDC typically also allow some extensions to SDC in order to achieve higher accuracy or better ease of use for the specific tool. Chapter 17 provides an overview of the Xilinx extensions to the SDC timing constraints – for their product Vivado™.

Conventions Used in This Book

In general, the names of SDC keywords and its options are printed in *italics*. *Italics* are also used to represent words that have a special meaning as it relates to this book.

Additional Resources

SDC is an open source format distributed by Synopsys, Inc. SDC Documentation and parsers can be downloaded for free from Synopsys website.

Feedback

We have put in our best efforts to provide an accurate description of the concepts. We also got help from some experts in the industry to review the material for accuracy. However, if you find some descriptions confusing or erroneous, please let us know.

Happy Reading!

Sanjay Churiwala
Sridhar Gangadharan

Acknowledgements

I would like to thank Mark Aaldering, who first showed me the need for a book on SDC. He explained: more and more tools were moving to SDC, and there was no book on the topic. With Xilinx also embracing SDC, there would be many users of Xilinx FPGAs, who would need to learn SDC.

I would also thank my friend Sridhar Gangadharan for his agreement to share the effort of writing the book.

Thanks are due to Charles Glaser of Springer. He readily accepted our proposal for this book.

I would like to dedicate this book to the people who shaped my understanding of the world of SDC and Timing Constraints – through numerous discussions on the topic of Timing Analysis during different stages of my career. The list is not exhaustive. However, some of the notable names include Fred Revenu (Xilinx), Greg Daughtry (Xilinx), Nupur Gupta (ST Microelectronics), Pankaj Jain (ST Microelectronics), Olivia Riewer (ST Microelectronics), K A Rajagopalan (Texas Instruments), Subrangshu (Shubro) Das (Texas Instruments), Satish Soman (Atrenta), Manish Goel (Atrenta), Shaker Sharwary (Atrenta) Pratyush Prasoon (Cadence), Sneh Saurabh (Cadence), Girjesh Soni (Synopsys), Rohan Gaddh (now, back to school), etc.

I would like to thank all my teachers, but, with a special mention to Prof. Swapna Banerjee of IIT Kharagpur, who introduced me to the world of VLSI during my days of Under-Graduation. In the early 1990s, Dhiraj Sogani (of Red Pine), Mithilesh Jha (Masamb), Ashutosh Varma (Cadence), and Late Dr. Narender Jain introduced me to the world of Timing. Then, towards the beginning of 2000, Sushil Gupta (Atrenta) and Vivek Gupta first introduced me to the world of SDC.

Thanks are due to Girjesh Soni (Synopsys), Subrangshu (Shubro) Das (Texas Instruments), Satish Soman (Atrenta), Shrinivasraj Muddey (Xilinx), Olivier Florent, Ravi Balachandran, and Russell Roan (Atrenta), for having reviewed portions of the book to ensure correctness of the material. I would also like to thank Fred Revenu (Xilinx) who wrote Chapter 17 of this book.

I would also like to thank Charu Puri for helping with all the illustrations in this book.

I would like to thank my management and the legal team at Xilinx for encouraging me to take on this activity. These include Mark Aaldering, Salil Raje, Vidya Rajagopalan, Scott Hover Smoot, and Sue Lynn Neoh. Ramine Roane (Xilinx) was also very encouraging. His only complain was: Why mid-2013? Why not earlier?

Special mention is always due to Mike Gianfagna (Atrenta) – whose involvement and association with any such activity is almost a sure-shot guarantee that all the barriers would be dealt with.

The confidence for writing this book came because of positive reaction to my previous book. The best reaction that I got was from Ramesh Dewangan (Atrenta) – who gifted a copy to his daughter; and from Prof. Ahmed Hemani (KTH Royal Institute of Technology, Sweden) – who prescribed the book to his students. In Indian context, this is one of the biggest respects one can show – when you show the confidence that my work is good enough to allow your children or students to learn from it. Thanks to them for showing this faith in me.

And, last but the most important one, we would remain indebted to our families and friends. Their cooperation and good wishes kept us going. And, the younger daughter (Lubha Churiwala – Std. VI) also helped me with sets of random reviews for Grammar and consistency:)

Hyderabad, India   Sanjay Churiwala

Acknowledgements

I had never imagined that I would author a book on Timing Constraints had it not been for an idea floated by Sanjay Churiwala. I would like to acknowledge a number of people who have helped me. My sincere thanks to each one of them because without their support, I couldn't have achieved this milestone.

First, I would like to thank God for this opportunity. Thanks to my parents for all the sacrifices they made to educate me and for being a constant source of inspiration and encouragement. I dedicate this book to them. I would also like to thank my wife Bulbul and my son Pratyush for supporting me in spite of all the time it took me away from them.

I would like to thank my mentor Dr. Ajoy Bose (CEO, Atrenta Inc.). Dr. Bose initiated me into the world of EDA 21 years ago and has been a guiding light for a significant part of my career. I would also like to thank my supervisor, Mo Movahed for his support and suggestions while I was writing this book. I would like to express my sincere gratitude to Mike Gianfagna for going out of his way to help me with contracts, proofreading, and providing access to all necessary resources.

I was introduced to the field of timing 2 decades ago by Ameesh Desai (LSI Logic). Thank you, Ameesh. I know you are up in heaven, smiling down on me.

I was very fortunate to have a stellar group of reviewers from the EDA and semiconductor industry who have helped in making the content of the book robust. I thank Girjesh Soni, Olivier Florent, Ravi Balachandran, Russell Roan, Satish Soman, Shrinivasraj Muddey, and Subrangshu Das for their candid feedback. I am grateful to Fred Revenu (Xilinx) for writing the chapter on Xilinx extensions. I would also like to thank Charu Puri for helping me with all the illustrations in this book.

I would like to acknowledge my dearest friend S. Raman (Managing Director, LSI India) for encouraging me at every step of my career.

I would like to thank my teachers at IIT Delhi who introduced me to the world of VLSI. Special credit is also due to my customers and colleagues (Amit Handa, Chandan Kumar, Hemant Ladhe, Irene Serre, Jean Philippe Binois, Manish Goel, Mark Silvestri, Neelu Bajaj, Nupur Gupta, Olivier Florent, Osamu Yaegashi, Pankaj Jain, Pratyush Prasoon, Shaker Sarwary, Sanjay Churiwala, Satish Soman,

Soenke Grimpen, Ramesh Dewangan, Russell Roan, Tanveer Singh, Tom Carlstedt-Duke, and Yutaka Tanigawa) who have helped me solidify my understanding of this domain. My apologies if I missed anyone.

Thanks to Sanjay for agreeing to partner with me on this book, and thanks to Springer for publishing this book.

Last but not the least, I would like to thank Synopsys, Inc. for coming up with a standard like SDC that has helped the design community immensely and for making it available as part of the tap-in program.

I hope you find this book useful.

San Jose, CA, USA   Sridhar Gangadharan

Contents

1	**Introduction**	1
	1.1 ASIC Design Flow	1
	1.2 FPGA Design Flow	4
	1.3 Timing Constraints in ASIC and FPGA Flow	7
	1.4 Timing Constraint Issues in Nanometer Design	7
	1.5 Conclusion	8
2	**Synthesis Basics**	9
	2.1 Synthesis Explained	9
	2.2 Role of Timing Constraints in Synthesis	10
	2.2.1 Optimization	10
	2.2.2 Input Reordering	11
	2.2.3 Input Buffering	12
	2.2.4 Output Buffering	12
	2.3 Commonly Faced Issues During Synthesis	13
	2.3.1 Design Partitioning	13
	2.3.2 Updating Constraints	14
	2.3.3 Multi-clock Designs	14
	2.4 Conclusion	15
3	**Timing Analysis and Constraints**	17
	3.1 Static Timing Analysis	17
	3.2 Role of Timing Constraints in STA	19
	3.2.1 Constraints as Statements	19
	3.2.2 Constraints as Assertions	20
	3.2.3 Constraints as Directives	20
	3.2.4 Constraints as Exceptions	21
	3.2.5 Changing Role of Constraints	21
	3.3 Common Issues in STA	22
	3.3.1 No Functionality Check	22
	3.3.2 No Check on Statements	22

	3.3.3	Requirements to be Just Right	23
	3.3.4	Common Errors in Constraints	24
	3.3.5	Characteristics of Good Constraints	24
3.4	Delay Calculation Versus STA		26
3.5	Timing Paths		26
	3.5.1	Start and End Points	26
	3.5.2	Path Breaking	28
	3.5.3	Functional Versus Timing Paths	28
	3.5.4	Clock and Data Paths	29
3.6	Setup and Hold		29
	3.6.1	Setup Analysis	29
	3.6.2	Hold Analysis	30
	3.6.3	Other Analysis	30
3.7	Slack		31
3.8	On-Chip Variation		32
3.9	Conclusion		33

4 SDC Extensions Through Tcl 35

4.1	History of Timing Constraints		35
4.2	Tcl Basics		36
	4.2.1	Tcl Variables	37
	4.2.2	Tcl Lists	38
	4.2.3	Tcl Expression and Operators	38
	4.2.4	Tcl Control Flow Statements	38
	4.2.5	Miscellaneous Tcl Commands	41
4.3	SDC Overview		41
	4.3.1	Constraints for Timing	42
	4.3.2	Constraints for Area and Power	42
	4.3.3	Constraints for Design Rules	42
	4.3.4	Constraints for Interfaces	43
	4.3.5	Constraints for Specific Modes and Configurations	43
	4.3.6	Exceptions to Design Constraints	43
	4.3.7	Miscellaneous Commands	44
4.4	Design Query in SDC		44
4.5	SDC as a Standard		44
4.6	Conclusion		46

5 Clocks 47

5.1	Clock Period and Frequency		47
5.2	Clock Edge and Duty Cycle		48
5.3	create_clock		50
	5.3.1	Specifying Clock Period	50
	5.3.2	Identifying the Clock Source	50
	5.3.3	Naming the Clock	51
	5.3.4	Specifying the Duty Cycle	51
	5.3.5	More than One Clock on the Same Source	52
	5.3.6	Commenting the Clocks	53

	5.4	Virtual Clocks	54
	5.5	Other Clock Characteristics	54
	5.6	Importance of Clock Specification	54
	5.7	Conclusion	55

6 Generated Clocks ... 57
- 6.1 Clock Divider ... 57
- 6.2 Clock Multiplier ... 58
- 6.3 Clock Gating ... 58
- 6.4 create_generated_clock ... 60
  - 6.4.1 Defining the Generated Clock Object ... 60
  - 6.4.2 Defining the Source of Generated Clock ... 60
  - 6.4.3 Naming the Clock ... 61
  - 6.4.4 Specifying the Generated Clock Characteristic ... 61
  - 6.4.5 Shifting the Edges ... 65
  - 6.4.6 More than One Clock on the Same Source ... 66
  - 6.4.7 Enabling Combinational Path ... 67
- 6.5 Generated Clock Gotchas ... 68
- 6.6 Conclusion ... 68

7 Clock Groups ... 71
- 7.1 Setup and Hold Timing Check ... 71
  - 7.1.1 Fast to Slow Clocks ... 73
  - 7.1.2 Slow to Fast Clocks ... 74
  - 7.1.3 Multiple Clocks Where Periods Synchronize in More than Two Cycles ... 74
  - 7.1.4 Asynchronous Clocks ... 75
- 7.2 Logically and Physically Exclusive Clocks ... 75
- 7.3 Crosstalk ... 76
- 7.4 set_clock_group ... 78
- 7.5 Clock Group Gotchas ... 80
- 7.6 Conclusion ... 80

8 Other Clock Characteristics ... 81
- 8.1 Transition Time ... 81
- 8.2 set_clock_transition ... 82
- 8.3 Skew and Jitter ... 83
- 8.4 set_clock_uncertainty ... 84
  - 8.4.1 Intraclock Uncertainty ... 84
  - 8.4.2 Interclock Uncertainty ... 85
- 8.5 Clock Latency ... 87
- 8.6 set_clock_latency ... 88
- 8.7 Clock Path Unateness ... 90
- 8.8 set_clock_sense ... 91
- 8.9 Ideal Network ... 93
- 8.10 Conclusion ... 94

9 Port Delays 95
9.1 Input Availability 95
9.1.1 Min and Max Availability Time 96
9.1.2 Multiple Clocks 97
9.1.3 Understanding Input Arrival Time 97
9.2 Output Requirement 98
9.2.1 Min and Max Required Time 99
9.2.2 Multiple Reference Events 100
9.2.3 Understanding Output Required Time 100
9.3 set_input_delay 101
9.3.1 Clock Specification 102
9.3.2 -level_sensitive 103
9.3.3 Rise/Fall Qualifiers 103
9.3.4 Min/Max Qualifiers 103
9.3.5 -add_delay 104
9.3.6 Clock Latency 105
9.3.7 Completing Input Delay Constraints 106
9.4 set_output_delay 106
9.4.1 Clock Specification 107
9.4.2 -level_sensitive 107
9.4.3 Rise/Fall Qualifiers 107
9.4.4 Min/Max Qualifiers 107
9.4.5 -add_delay 108
9.4.6 Clock Latency 108
9.4.7 Completing Output Delay Constraints 108
9.5 Relationship Among Input and Output Delay 108
9.6 Example Timing Analysis 110
9.6.1 Input Delay: Max 110
9.6.2 Input Delay: Min 111
9.6.3 Output Delay: Max 112
9.6.4 Output Delay: Min 113
9.7 Negative Delays 114
9.8 Conclusion 114

10 Completing Port Constraints 117
10.1 Drive Strength 117
10.1.1 set_drive 118
10.2 Driving Cell 119
10.2.1 set_driving_cell 120
10.3 Input Transition 125
10.3.1 Input Transition Versus Clock Transition 125
10.4 Fanout Number 126
10.5 Fanout Load 126
10.6 Load 127
10.6.1 Net Capacitance 127
10.6.2 Pin Load Adjustments 128

		10.6.3	Load Type	128
		10.6.4	Load Versus Fanout Load	129
		10.6.5	Load at Input	129
	10.7	Conclusion		129

11 False Paths 131
- 11.1 Introduction 131
- 11.2 set_false_path 131
- 11.3 Path Specification 132
- 11.4 Transition Specification 135
- 11.5 Setup/Hold Specification 137
- 11.6 Types of False Paths 137
  - 11.6.1 Combinational False Path 137
  - 11.6.2 Sequential False Path 138
  - 11.6.3 Dynamically Sensitized False Path 139
  - 11.6.4 Timing False Path 139
  - 11.6.5 False Path Due to Bus Protocol 140
  - 11.6.6 False Path Between Virtual and Real Clocks 141
- 11.7 set_disable_timing 142
- 11.8 False Path Gotchas 143
- 11.9 Conclusion 144

12 Multi Cycle Paths 145
- 12.1 SDC Command for Multi Cycle Paths 145
- 12.2 Path and Transition Specification 146
- 12.3 Setup/Hold Specification 147
- 12.4 Shift Amount 148
- 12.5 Example Multi Cycle Specification 151
  - 12.5.1 FSM-Based Data Transfer 151
  - 12.5.2 Source Synchronous Interface 152
  - 12.5.3 Reset 154
  - 12.5.4 Asynchronous Clocks 154
  - 12.5.5 Large Data Path Macros 154
  - 12.5.6 Multimode 155
- 12.6 Conclusion 155

13 Combinational Paths 157
- 13.1 set_max_delay 157
- 13.2 set_min_delay 158
- 13.3 Input/Output Delay 158
  - 13.3.1 Constraining with Unrelated Clock 159
  - 13.3.2 Constraining with Virtual Clock 159
  - 13.3.3 Constraining with Related Clock 160
- 13.4 Min/Max Delay Versus Input/Output Delay 161
- 13.5 Feedthroughs 162
  - 13.5.1 Feedthroughs Constrained Imperfectly 164

	13.6	Point-to-Point Exception	164	
	13.7	Path Breaking	165	
	13.8	Conclusion	166	
14	**Modal Analysis**		**167**	
	14.1	Usage Modes	167	
	14.2	Multiple Modes	167	
	14.3	Single Mode Versus Merged Mode	169	
	14.4	Setting Mode	169	
	14.5	Other Constraints	172	
	14.6	Mode Analysis Challenges	172	
		14.6.1 Timing Closure Iterations	172	
		14.6.2 Missed Timing Paths	173	
	14.7	Conflicting Modes	173	
	14.8	Mode Names	175	
	14.9	Conclusion	175	
15	**Managing Your Constraints**		**177**	
	15.1	Top-Down Methodology	177	
	15.2	Bottom-Up Methodology	178	
	15.3	Bottom-Up Top-Down (Hybrid) Methodology	181	
	15.4	Multimode Merge	183	
		15.4.1 Picking Pessimistic Clock	185	
		15.4.2 Mutually Exclusive Clocks	185	
		15.4.3 Partially Exclusive Clocks	186	
		15.4.4 Merging Functional and Test Mode	188	
		15.4.5 Merging I/O Delays for Same Clock	189	
		15.4.6 Merging I/O Delays with Different Clocks	189	
	15.5	Challenges in Managing the Constraints	190	
	15.6	Conclusion	192	
16	**Miscellaneous SDC Commands**		**193**	
	16.1	Operating Condition	193	
		16.1.1 Multiple Analysis Conditions	195	
		16.1.2 set_operating_conditions	195	
		16.1.3 Derating	196	
	16.2	Units	197	
	16.3	Hierarchy Separator	198	
		16.3.1 set_hierarchy_separator	198	
		16.3.2 -hsc	199	
	16.4	Scope of Design	200	
		16.4.1 current_instance	200	
	16.5	Wire Load Models	201	
		16.5.1 Minimal Size for Wire Load	202	
		16.5.2 Wire Load Mode	202	
		16.5.3 Wire Load Selection Group	203	

	16.6	Area Constraints	204
	16.7	Power Constraints	205
		16.7.1 Voltage Island	205
		16.7.2 Level Shifters	206
		16.7.3 Power Targets	207
	16.8	Conclusion	207
17	**XDC: Xilinx Extensions to SDC**		**209**
	17.1	Clocks	209
		17.1.1 Primary and Virtual Clocks	210
		17.1.2 Generated Clocks	210
		17.1.3 Querying Clocks	212
		17.1.4 Clock Groups	213
		17.1.5 Propagated Clocks and Latency	214
		17.1.6 Clock Uncertainty	215
	17.2	Timing Exceptions	216
	17.3	Placement Constraints	217
	17.4	SDC Integration in Xilinx Tcl Shell	218
	17.5	Conclusion	218
Bibliography			**219**
Index			**221**

List of Figures

Fig. 1.1	A sample circuit for scan insertion	3
Fig. 1.2	ASIC design flow	4
Fig. 1.3	Internal representation of an FPGA	5
Fig. 1.4	Representative logic block	5
Fig. 1.5	Switch box	6
Fig. 1.6	FPGA design flow	6
Fig. 2.1	ANDing of 4 inputs	11
Fig. 2.2	Alternative realization of Fig. 2.1	11
Fig. 2.3	Input being buffered	12
Fig. 2.4	A design partitioned into blocks	13
Fig. 3.1	A sample (abstract) circuit for STA basics	18
Fig. 3.2	Internals of the circuit shown in Fig. 3.1	18
Fig. 3.3	Sample statement-type constraints	19
Fig. 3.4	Clock tree	20
Fig. 3.5	Output required time constraint	21
Fig. 3.6	Constraints accuracy requirements	23
Fig. 3.7	Intent of constraints	25
Fig. 3.8	Exceptions being invalidated due to RTL change	25
Fig. 3.9	Timing paths	27
Fig. 3.10	Setup check at an output	29
Fig. 3.11	Slack	31
Fig. 3.12	On-chip variation	32
Fig. 5.1	Clock waveform	48
Fig. 5.2	Clock waveform with uneven duty cycle	49
Fig. 5.3	Positive edge-triggered circuit	49
Fig. 5.4	Negative edge-triggered circuit	49
Fig. 5.5	Clock waveform with low pulse	51
Fig. 5.6	Clock with complex waveform	52
Fig. 5.7	Block driven by off-chip multiplexer with two clocks	53

Fig. 6.1	2-bit ripple counter	58
Fig. 6.2	A simple clock multiplier	59
Fig. 6.3	A gated clock	59
Fig. 6.4	A gated clock to generate pulse	59
Fig. 6.5	Divide-by-two circuit with a non-inverting clock	63
Fig. 6.6	Divide-by-two circuit with an inverting clock	63
Fig. 6.7	Block driven by off-chip multiplexer with two clocks	66
Fig. 6.8	Source-synchronous interface	67
Fig. 7.1	Asynchronous clock domain crossing	72
Fig. 7.2	Waveform for interacting clocks	72
Fig. 7.3	Waveform for fast to slow clocks	73
Fig. 7.4	Waveform for slow to fast clocks	74
Fig. 7.5	Waveform for clocks that are not integer multiples	74
Fig. 7.6	Logically exclusive clocks (*C1* and *C2*)	75
Fig. 7.7	Physically exclusive clocks (*GC1* and *GC2*)	76
Fig. 7.8	Glitch due to crosstalk	77
Fig. 7.9	Victim slew deterioration on account of crosstalk	77
Fig. 8.1	Waveform of a non-ideal clock	82
Fig. 8.2	Impact of uncertainty on setup and hold	84
Fig. 8.3	C1 to C2 and C2 to C1 paths	86
Fig. 8.4	(a) Delay on clock path with off-chip clock source (b) Delay on clock path on-chip clock source	87
Fig. 8.5	Setup and hold check with source latency	89
Fig. 8.6	Positive and negative unate clock	91
Fig. 8.7	Non-unate clock	91
Fig. 8.8	Different kinds of pulse waveforms	93
Fig. 9.1	Block input	96
Fig. 9.2	Input available time	96
Fig. 9.3	Multiple paths from same reference event	96
Fig. 9.4	Data valid window	97
Fig. 9.5	Multiple reference events	98
Fig. 9.6	Block output	99
Fig. 9.7	Output required time	99
Fig. 9.8	Multiple paths to same reference event	100
Fig. 9.9	Multiple reference events	101
Fig. 9.10	Clock specification for input delay	102
Fig. 9.11	Clock latency impact on input delay	105
Fig. 9.12	Input and output delay relation	109
Fig. 10.1	Equivalent resistance	118
Fig. 10.2	NAND driver	119
Fig. 10.3	Driver cell	120
Fig. 10.4	Driver with multiple loads	122

List of Figures   xxv

Fig. 10.5	Clock specification for driving cell	124
Fig. 10.6	Clock network	126
Fig. 10.7	Fanout load	127
Fig. 10.8	Pin load adjustments for net capacitance	128
Fig. 10.9	Slew degradation through wire	129
Fig. 11.1	A circuit represented as a graph network	132
Fig. 11.2	Clock network with non-inverting paths but registers triggered by both edges	135
Fig. 11.3	Recovery and removal timing check	137
Fig. 11.4	Combinational false path	138
Fig. 11.5	Sequential false path	138
Fig. 11.6	Dynamically sensitized false path	139
Fig. 11.7	Protocol-based data exchange	140
Fig. 11.8	Multiplexed output pin	141
Fig. 11.9	Combinational loop	142
Fig. 12.1	Default setup timing relationship	146
Fig. 12.2	Multi cycle of 2	146
Fig. 12.3	Clock waveform	147
Fig. 12.4	Start and end clocks have different period	149
Fig. 12.5	Start clock slower than end clock	150
Fig. 12.6	FSM-based data transfer	151
Fig. 12.7	Simple realization of source synchronous interface	152
Fig. 12.8	Source synchronous interface – corresponding waveform	152
Fig. 12.9	Asynchronous clocks	154
Fig. 13.1	Combinational path	158
Fig. 13.2	Combinational path – no interaction with clock	159
Fig. 13.3	Combinational path in the context of launching/capturing flop	160
Fig. 13.4	An input/output is part of combinational as well as registered path	161
Fig. 13.5	Feedthrough path spanning multiple blocks	163
Fig. 13.6	Simple double-flop synchronization	164
Fig. 13.7	Path breaking	165
Fig. 14.1	Functional and test mode	168
Fig. 14.2	Scan pins shown for the previous circuit	170
Fig. 14.3	Case analysis impact on paths being timed	171
Fig. 14.4	Conflicting mode settings	174
Fig. 15.1	Bottom-up constraints propagation causing conflict	178
Fig. 15.2	Bottom-up constraints propagation	179
Fig. 15.3	Chip vs. block constraints validation	180
Fig. 15.4	Budgeting chip constraints	182
Fig. 15.5	Picking pessimistic clock	185

Fig. 15.6	Mutually exclusive clocks	186
Fig. 15.7	Partially exclusive clocks	187
Fig. 15.8	Design with scan chain hooked up	188
Fig. 15.9	Design with false path before optimization	191
Fig. 15.10	Design after optimization	192
Fig. 16.1	Object reference through current_instance	200
Fig. 16.2	Wire spanning across multiple hierarchies	203
Fig. 17.1	Tool-created generated clock on a Xilinx PLL output	210
Fig. 17.2	Getting a clock along its tree	212
Fig. 17.3	Getting a clock and its associated generated clocks	213
Fig. 17.4	Asynchronous clocks	214

List of Tables

Table 4.1	List of supported operators in Tcl	39
Table 4.2	Commonly use Tcl commands	41
Table 4.3	Constraints for timing	42
Table 4.4	Constraints for area and power	43
Table 4.5	Constraints for design rules	43
Table 4.6	Constraints for interfaces	43
Table 4.7	Constraints for specific modes and configurations	44
Table 4.8	Exceptions to design requirements	44
Table 4.9	Miscellaneous commands	44
Table 4.10	Commands for design query	45

Chapter 1
Introduction

Application-specific integrated circuit (ASIC) is an IC targeted for a specific application, e.g., chips designed to run graphics on a game console, standard interfaces like USB, PCI bus to consumer electronics, special functions to control automotive electronics, and chips for smart phones. In the early days of chip design, ASICs were a few thousand gates. With advancements in deep submicron technology, today's ASICs run into millions of gates. Today, some of the more complex ASICs combine processors, memory blocks, and other ASIC or ASIC derivatives called IPs (intellectual property). These are called SoCs or system on a chip. The reality is today's SoCs will become the blocks or IPs for the future SoC design. This complex nature of ASIC development requires a well-structured design flow that is scalable and provides enough flexibility to designers and SoC integrators alike to define a methodology for seamless design.

Another paradigm for IC design that has gained popularity in recent years is FPGA (field-programmable gate arrays). FPGAs can be used to implement any function that can be developed as an ASIC, the only difference being that an IC designed as an FPGA can be programmed by the user after its manufacturing. This ability to field-program the IC doesn't restrict the user to any predetermined hardware function, and IC can be tweaked based on changing standards providing reduced nonrecurring engineering cost and significant time to market advantages over ASICs, although taking a hit on the performance and power consumption.

This chapter provides an overview to the typical design flow in ASIC and FPGA design. It also covers how timing constraints impact these flows.

1.1 ASIC Design Flow

A typical ASIC flow can be broadly categorized into logical design and physical design. Logical design begins with high-level design specification and chip architecture. The chip architect captures high-level functional, power (how much power should the design consume) and timing (at what speed should the design operate) requirements. This is followed by a register transfer-level description of the design. Commonly referred to as RTL (register transfer level), this provides an abstraction of the functional behavior of the circuit in terms of how the logical operations on signals enable data to flow between registers (flops) in a design. This is typically captured using hardware description languages (also referred to as HDLs) like Verilog, SystemVerilog, and VHDL. Once the functionality of the design is coded, it is verified using simulation. Simulation is a process where various stimuli are applied to a representation of a design and the response of the design is captured. The objective is to validate that the resulting output matches the desired functionality of the circuit. For example, if you implement an adder, which includes two inputs and one output, the test vector will emulate inputs as two numbers that need to be added, and output should represent the sum of these numbers. At this stage, the design is ready for synthesis.

Synthesis (aka logic synthesis) is the step where RTL description is translated to a gate-level representation which is a hardware equivalent implementation of the functionality described in the HDL. Let us consider the following description in Verilog:

module flipflop (d, clk, rst, q)
input d, clk, rst;
output q;

reg q;
always @(posedge clk)
if (rst)
   q <= 1'b0;
else
   q<= d;
endmodule

Synthesis tool will map this RTL description to a positive-edge-triggered flip-flop with a synchronous reset. An HDL description is said to be synthesizable RTL, if it can be consumed by industry standard synthesis tools to map to a unique and unambiguous implementation. In this step, the designer also captures certain design and timing characteristics which are representative of the high-level objectives set forth by the chip architect, like clock frequency, delays available in the block, and target library, so that the synthesis tool can optimize the design to meet the requirements. Details on synthesis are available in Chap. 2 of this book.

After synthesis, the design is prepared for testability. DFT or design for testability is the technique to ensure that there are enough hooks in place to perform tests on the IC after manufacturing so that faulty parts don't get shipped. One such technique is called scan insertion, also known as test-point insertion. Let us consider the circuit in Fig. 1.1.

In this circuit the second flop is not controllable. However, by inserting the multiplexer structure, the user can control the second flop via a *scan clock* and *scan enable*. This results in all registers to be linked in a chain called the scan chain or the scan path. Similar to clock control, the data to the flop can also be controlled using the scan enable. This is used to test that data can be shifted through the design. This technique helps to make all registers in the design controllable and observable via the inputs and outputs of the design.

1.1 ASIC Design Flow

Fig. 1.1 A sample circuit for scan insertion

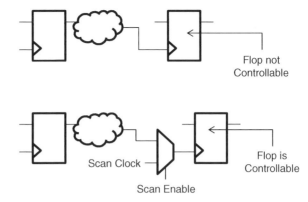

After synthesis and scan insertion, the hardware equivalent representation needs to be verified against the original RTL description so that the design intent is preserved. This is called *equivalence checking* and uses *formal verification* techniques. At this stage design is also ready for *STA* or *static timing analysis*. It is worthwhile to note that equivalence checking only verifies the functionality of the implemented gate-level representation against the original description but not whether it meets the frequency target of the implementation, which is the responsibility of STA.

STA is a method of checking the ability of the design to meet the intended timing requirements, statically without the need for simulation. Most STA engines require the designers to specify timing constraints that model how the chip needs to be characterized at its periphery and what assumption to make inside the design so as to meet the timing requirements set forth by the chip architect. This is specified using an industry standard format called SDC (Synopsys Design Constraints) which forms the premise of this book. Details on STA are available in Chap. 3 onwards of this book. STA step completes the logical design step and acts as the bridge between logical and physical design.

Physical design begins with floor planning. After preliminary timing analysis, the logical blocks of the design are placed with the aim of optimizing area, aspect ratio, and how signals between the blocks interact. The objective is to ensure that there isn't too much of inter-block interaction that causes congestion or difficulty in routing. These factors have direct impact on power, area, timing, and performance. Once the optimal floor plan is achieved, the connections between blocks are routed. During the synthesis stage, many assumptions are made about clock network since that level of design information is not available until after floor plan. Floor planning is followed by clock tree synthesis to distribute the clock as evenly as possible so as to reduce clock skew between different parts of the design. This step of floor planning, placement, and routing is called *layout* of a design. During the physical design, STA may be done multiple times to perform a more accurate timing analysis as the assumptions made during the initial implementation are gradually solidified.

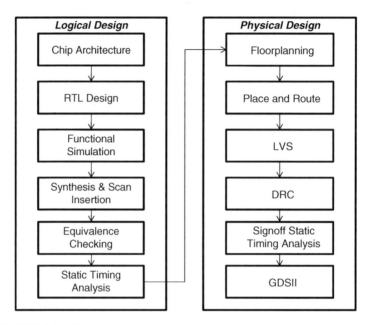

Fig. 1.2 ASIC design flow

At this stage layout of an IC needs to be verified to ensure the following criteria are met:

1. All rules laid out by the foundry where it will be fabricated into a chip are adhered to. This is called *DRC* or Design Rule Checking.
2. The layout matches the netlist generated after synthesis. This is called *LVS* or layout versus schematic where the layout is formally verified against the post-synthesis netlist.

Once the design is *LVS* and *DRC* clean, signoff static timing analysis is completed. After layout, design is not guaranteed to meet timing and may require to be tweaked further so as to meet the timing and frequency requirements. After signoff static timing analysis is successful, GDSII of the design is generated. *GDSII* is a geometric representation of the polygons that describe the actual layout of the design with all its connectivity. Fabs manufacture chips based on the *GDSII* that is released to them.

This whole flow from logical to physical design is commonly referred to as the *RTL2GDSII* flow, and process of releasing GDSII for manufacturing is termed as *tapeout*. Figure 1.2 shows the typical ASIC design flow.

1.2 FPGA Design Flow

FPGA comprises an array of logical blocks and connections between blocks, both of which are programmable. Figure 1.3 shows the internal representation of an FPGA.

1.2 FPGA Design Flow

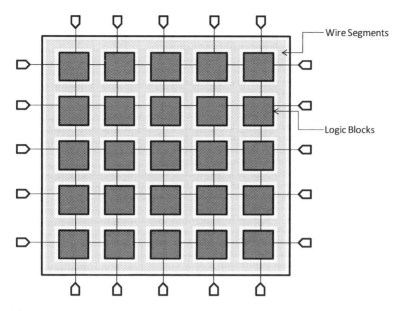

Fig. 1.3 Internal representation of an FPGA

Fig. 1.4 Representative logic block

Logic block is used to define the functionality of the circuit, and its complexity can vary depending on the function the user is trying to implement or the target FPGA from the vendor. This has direct correlation to the placement, routing, and timing analysis. Figure 1.4 shows a representative hypothetical logic block structure.

The logic block consists of a *LUT* or lookup table, which can be used to implement any arbitrary combinational function. The output of the LUT is then registered or connected to the output directly. *Tracks*, a collection of horizontal and vertical wire segments, run between the logical blocks. These can be programmed using switch boxes to indicate the actual connectivity between the intersecting horizontal and vertical wires. Figure 1.5 shows the internals of a switch box.

FPGA design flow is similar to ASIC flow in the logical design portion. The user writes the RTL description using one of the HDLs. The HDL is functionally verified using simulation and then synthesized to logic gates. However, the physical design is vendor dependent. Post synthesis, the netlist is compiled to target FPGA on which it needs to be mapped. This compilation step includes mapping netlist functionality to logic block, placing the logic blocks and routing between the blocks using the tracks available in the target FPGA. Place and route is timing constraints driven to ensure the

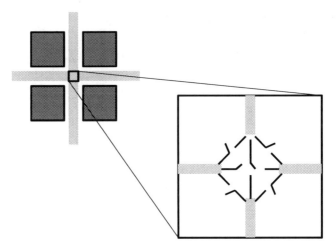

Fig. 1.5 Switch box

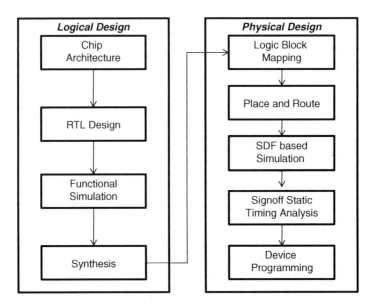

Fig. 1.6 FPGA design flow

timing requirements of the IC are met. Once place and route is complete, the delays of the circuit are generated in *SDF* (standard delay format). This SDF file is used along with the post-layout netlist to do back-annotated full-timing gate-level simulation (*FTGS*). Since simulation is not always exhaustive, accurate static timing analysis is also performed at this stage. Once all the verification is complete, the device is programmed. Figure 1.6 shows the FPGA design flow.

Since FPGA flow is faster to execute, it has now become quite common to prototype ASICs and SoCs using an FPGA. This is called FPGA prototyping.

1.3 Timing Constraints in ASIC and FPGA Flow

The successful tapeout of any chip is measured by a variety of factors. This includes how well the design adheres to the timing, power, and area objectives set by the architect in addition to meeting all the functional requirements. Given the complexity of both ASIC and FPGA design flows, it is prudent to establish checks and balances at each stage of the design flow for this measure to prevent any late-stage design changes and ECOs.

From a timing perspective, at the architecture stage the architect will assign block budgets which are handed off to block owners. Depending on whether a block is a derivative design or being developed from scratch, the RTL designer will create initial timing constraints or tweak existing ones for synthesis. This will form the baseline for all runs in the implementation flow and typically includes defining clock frequency and budgets in the subblocks. This results in an unoptimized netlist with ideal clocks (clock with zero delay). Once logic optimization step is completed by the synthesis tool, STA is done. At this stage more accurate timing intent in the form of intra-block delay, clock latency, and clock skew is provided, with objective that design meets all setup and hold requirements and correctly estimates any interconnect delay.

During the physical design stage clock assumptions (skew and network delays) made during logical design get solidified. Delays can be computed more accurately using the actual parasitics extracted from the real routing. Clock tree synthesis is done to balance the clock tree to reduce any clock skew.

Timing is critical component in this flow and its impact is uniform whether in the ASIC or FPGA flow. If you look at the evolution of chip, timing plays an integral part at each step of the flow. It is constantly tweaked and verified as the design progresses through the implementation flow. At each step the designer tries to ensure that original timing intent as prescribed by the chip architect is preserved. A significant duration of physical design cycle is spent on achieving timing closure.

1.4 Timing Constraint Issues in Nanometer Design

As described in the last section, timing constraints touch many stages of design flow. Given the strong dependence, designers face many pain points ranging from creation, verification, to validation of timing constraints. During the logical design step, creation of constraints is an error-prone and iterative process. The designer needs to translate the chip architect's requirements into equivalent, yet reasonable, constraints. However, at this stage most designers are concerned about meeting the functional requirements, and timing ends up taking a backseat. Once block-level constraints are created, they need to be validated for correctness. The designers also need to ensure that constraints are in sync with the design.

When block design is complete, the design and the constraints are handed off to the subsystem or chip integrator. The integrators develop their own top-level constraints which have to be consistent with the blocks; otherwise, a block which may have met timing may not function when integrated into the subsystem or chip.

During the physical design, power optimization changes like clock gating may not take into account its impact on timing constraints, thereby deviating from the original timing intent. To aggravate problems further, there is knowledge disconnect between RTL design and physical design teams. RTL design teams tend to be less concerned about timing since their objective is to meet the functionality. The physical design team is responsible for meeting the timing; however, they don't know the internals of the design. They have to rely on RTL designer to bless the timing constraints. This disconnect results in unnecessary iterations having direct impact on time to market.

1.5 Conclusion

ASIC and FPGA design flows depend heavily on timing constraints. Most design flows are heavily focused on verifying the correctness of the functionality of the design – represented through RTL. However, an equal emphasis has traditionally not been given to validation of timing constraints. Timing constraints issue can cause unpredictable design schedule, delay tapeout, increase iterations between logical and physical design teams, and result in late-stage ECOs. A set of constraints if not written properly can greatly diminish the ability to reuse the block in future SoCs. Since constraints impact performance of the realized hardware design, the quality of timing constraints has a direct correlation to the quality of silicon. As we will see in subsequent chapters of this book, the world of constraints is full of very fine-grained nuances. Thus, it is very important to understand and write constraints which are correct as well as efficient.

Chapter 2
Synthesis Basics

Synthesis is the first step in the design process, where timing constraints are used.

2.1 Synthesis Explained

Let us consider a 3-bit counter, which counts in the sequence $0 \rightarrow 5 \rightarrow 2 \rightarrow 7 \rightarrow 6 \rightarrow 3 \rightarrow 5 \rightarrow 1 \rightarrow 0$. If we have to realize the gate-level circuit for this counter, it would take a lot of time to draw the *Karnaugh map* and then realize the logic feeding into the *D* pin of each of the 3 flops which form the counter.

However, it is much faster to write an HDL code, which describes the above functionality. This HDL code can then be taken through a tool, which will create the corresponding netlist.

Synthesis in the context of electronic design means realization of a gate-level netlist to achieve a specific functionality. Besides the specific functionality, the process of synthesis might also meet certain other requirements, namely, power, frequency of operation, etc.

Sometimes, there are specialized synthesis tools for specific kinds or portions of circuit, e.g.:

- Clock tree synthesis – which creates the clock tree
- Data path synthesis – which creates a repetitive structure in the data path
- Logical synthesis – used for realizing all kinds of logical circuits

  Usually, the word "synthesis" just by itself means logical synthesis only.

2.2 Role of Timing Constraints in Synthesis

The design process involves a lot of steps. These steps are of various kinds, e.g.:

- Capturing intent
- Verifying that the design is in line with what we desire
- Estimating certain characteristics
- Actually realizing the design

The last series of steps are also called implementation steps. Synthesis is the first among the implementation steps. The following subsections give a few examples of the choices that a synthesis tool might need to make and the basis of the decision. These are all examples of additional information (beyond functionality) that the synthesis tool needs to be provided through constraints.

2.2.1 Optimization

For a synthesis tool to realize a netlist, it needs several pieces of information. The first information is the functionality that the realized netlist needs to perform. This information comes from the HDL description.

For a device, obviously functionality is the most important consideration. However, designers have to be also very sensitive to:

- Area: We want to fit as much functionality into the same unit area as possible.
- Power: We want to conserve battery power and also reduce junction heating.
- Performance: We want to get highest possible speed from the device.

However, each of the above goals may impact the others and sometimes negatively. For example, if we want to get best speed, we will need to have higher drive devices, which will mean higher power and greater area on silicon. So, instead of a designer trying to squeeze out the maximum performance, the designer might want to get just about enough performance that would achieve the purpose and, in the process, save on area and power.

A designer communicates his requirements around area, power, and performance to the synthesis tool through *constraints*. Once the synthesis tool is able to achieve a circuit that meets these goals, the tool need not make any further effort to realize a "better" circuit. Any further attempt to improve in any one dimension could worsen the other dimensions.

So constraints are used to tell the synthesis tool – among the many possible implementations possible to realize the same functionality, which should be chosen so that all the three requirements on area, power, and performance are met.

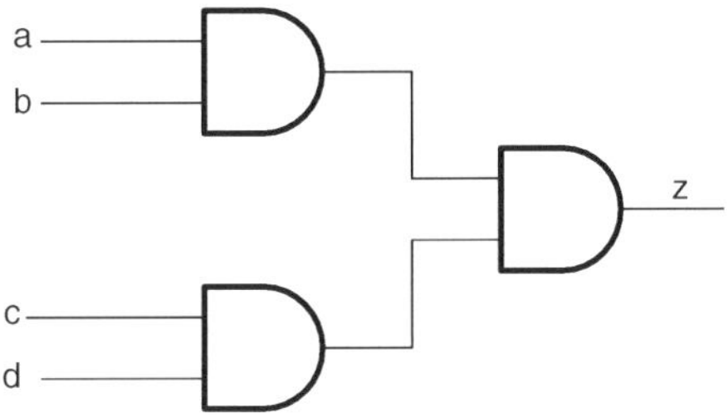

2.2 Role of Timing Constraints in Synthesis

Fig. 2.1 ANDing of 4 inputs

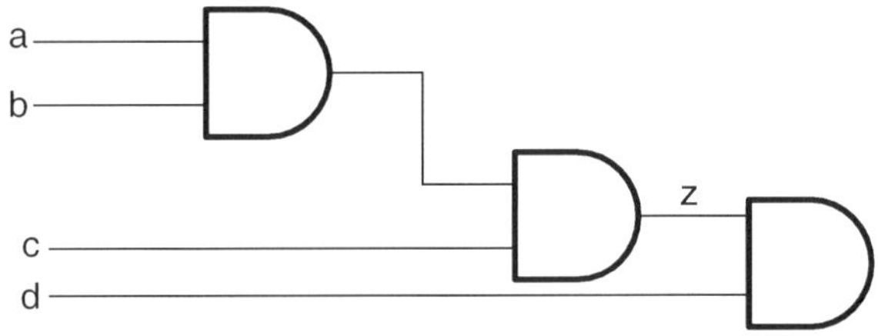

Fig. 2.2 Alternative realization of Fig. 2.1

2.2.2 Input Reordering

Let us consider a function involving *AND*ing of four inputs, a, b, c, and d. One of the simplest realizations of this circuit is as shown in Fig. 2.1.

However, now imagine that the input d arrives much later than other inputs. So the final evaluation of the circuit has to wait till d arrives and passes through 2 *AND* gates. On the other hand, there can be an alternative realization of the same functionality as shown in Fig. 2.2.

In this circuit, by the time d arrives, the other three signals have already been evaluated, and d has to travel through only one *AND* gate.

Though both circuits perform the same functionality and have similar area (3 *AND* gates), a designer might prefer Fig. 2.1 or 2.2, depending upon whether d comes along with all other signals or whether d comes much later than all other signals. If instead of d, it was some other signal which was coming much later, then d might be swapped with that late arriving signal.

Thus, depending upon the relative arrival time for various inputs feeding into the same combinational logic, the synthesis tool might need to decide which design should be chosen among the available choices – so that the last arriving signals have to cross the minimum number of logic.

Designers use constraints to convey to the synthesis tool about the arrival time of various input signals.

Fig. 2.3 Input being buffered

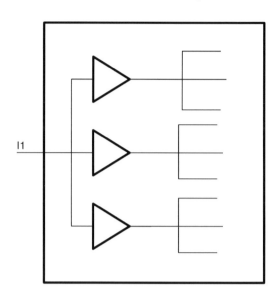

2.2.3 Input Buffering

Drive can be thought of as current-carrying capability. Thus, higher drive means output would switch faster and a higher amount of load can be connected. Let us say, a specific input has to drive a huge fanout cone. But, whether the specific input can drive such a huge cone or not depends upon the driving capability of the signal which is driving the input. If the signal driving the input cannot drive the load for the whole fanout cone, then the signal would need to be buffered before it can be fed into the huge cone.

Figure 2.3 shows an input which has to drive a fanout load of *9*. However, it does not have the drive strength for that kind of load. Hence, buffering is done on the input, before feeding into the load. With this buffering, the load that the input has to drive is only *3*.

Designers need to tell the synthesis tool the driving capability of the external signal which is driving the input so that synthesis tool can decide whether or not to put additional buffers. And constraints are used to convey information about the drive strength of the external inputs.

2.2.4 Output Buffering

Similar to input buffering, a design might need to have additional drive capability at the output side, if the output port is expected to drive a huge load externally. So designers need to convey to the synthesis tool – the external load that a port

2.3 Commonly Faced Issues During Synthesis

might have to drive. Synthesis tool will then choose appropriate cells or buffers with the right drive strengths that can drive the load. And constraints are used to convey information about the external load that needs to be driven by the output port.

2.3 Commonly Faced Issues During Synthesis

Synthesis step can have different class of issues. In fact, one could write a whole book around issues faced during synthesis. This section gives a glimpse of some issues around synthesis related to constraints. These same topics are discussed in much more details in subsequent chapters of the book.

2.3.1 Design Partitioning

Though synthesis techniques have provided a major leap in terms of designer's productivity, the biggest bottleneck of a synthesis tool is the size of a design that it can synthesize. The design sizes today are humongous, compared to the sizes of design that synthesis tool can synthesize.

Thus, a full design has to be broken into smaller units, called blocks. During synthesis stage, the blocks are created based on logical view of the design, namely, related functionality being put into one block. This kind of partitioning is called *logical partition*. A synthesis tool would synthesize one block as a unit. Thus, a synthesis tool can view only a block at any given time, and it does not see how the block interacts with the rest of the design. Figure 2.4 shows how a design is composed of logical blocks.

The outermost rectangular boundary denotes the complete design. Usually, the design would have requirements listed for the whole design. Because the synthesis tool cannot synthesize the whole design, so the design is partitioned into smaller blocks (*B1* through *B6*), represented by inner smaller rectangles.

At any time, synthesis tool views a block. But, the requirements are known for the complete design. So the top-level constraints for the complete design have to be

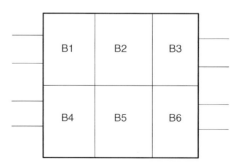

Fig. 2.4 A design partitioned into blocks

broken into constraints for individual blocks. These constraints for individual blocks have to be created – based on interaction of the block with all other blocks. For example, for the block *B1*, the constraints have to be specified to define its interaction with primary inputs for the design as well as its interaction with other blocks *B2* and *B4*.

So what was supposed to be just the constraints at the top level now gets translated into many more constraints defined at each interface. And as the number of constraints grows, there are higher chances of errors. In the figure, the partitions are shown as regular rectangular blocks. In reality, all the blocks interact with many blocks, and that increases the complexity of the total set of constraints.

Let us consider the interaction between blocks *B1* and *B2*. Based on this interaction, there would be some constraints for blocks *B1* and corresponding constraints for block *B2*. Many times, the people or the team working on these different blocks are different. There have been many instances where the constraints written for interfacing blocks are not consistent. For example, *B1*'s designer might assume that he will get 50 % of the total path time for his block and the remaining 50 % would be for rest of the path. Similarly, *B2*'s designers might also assume 50 % of the path time available for his block. So between the two blocks, they might consume the entire path time, leaving nothing for the top-level routing for connecting the two blocks.

2.3.2 Updating Constraints

It seems slightly strange that such inconsistency might happen among blocks of the same design. However, such inconsistencies usually creep in gradually as various blocks keep getting impacted due to some other block not meeting their initial requirements.

Let us assume block *B1* failed to meet some of its timing, which causes an impact on *B2*. Block *B2*'s designer might now have to update his constraints, and its impact might be on the *B2/B3* and *B2/B5* interface. However, at this stage, either *B3* or *B5* constraints might get out of sync with the updated constraints of *B2*, and in many cases, these changed constraints might disturb delicate balance of area, performance, and power. Thus, the block-level constraints may have to be updated depending on how the block is integrated in the subsystem or chip.

2.3.3 Multi-clock Designs

Most designs today have multiple processing cores, running on different clock frequencies. There could be different peripherals for these cores. In the process of integrating these cores which are being developed simultaneously by design groups, an inadvertent mistake of constraining a high-frequency core with a low-frequency constraint may be missed during initial implementation. These may be eventually caught during full-chip STA, post integration. That could be pretty late as the block constraints would now have to be redone to the original specification adding an unnecessary iteration to the chip integration.

2.4 Conclusion

This chapter gave a glimpse of the need for constraints and nature of some of the problems related to synthesis. Synthesis has been used just as an example of an implementation tool. All implementation tools are driven by constraints. Most of the discussions mentioned in this chapter would apply to all implementation tools, not just synthesis. So incorrect constraints impact the ability of these tools to implement a circuit which will meet its performance, area, and power goals.

As design complexities are growing, the constraints themselves are also becoming complex – in order to be able to correctly represent the complex requirements as well as relationships. The nuances of the design process involving partitioning, integration, and multiple cores operating at different frequencies all add to further problems around creating constraints.

Several implementation tools also allow constraints to provide physical information, such as physical shape of a block, or specific location of ports, etc. These physical constraints are not covered in this book.

Chapter 3
Timing Analysis and Constraints

Before we learn to constrain our design, let us first understand the basics of timing analysis. Fundamentally, timing analysis is of two kinds:

- Dynamic timing analysis
- Static timing analysis (STA)

Dynamic timing analysis means we apply a set of vectors at the inputs and observe the time at which the signals reach various points in the circuit. By knowing the time difference between the inputs applied and the signals observed, we know how long the signal takes to travel through the specific path segment. For flops, by observing when *D* input arrives with respect to the *CLK* input, we know whether or not the specific flop meets the setup and the hold requirements. Thus, this process is dependent on timing simulation and is dependent on the stimulus being applied. On the other hand, static timing analysis analyzes the circuit topology to compute the same information without the need for any input vectors.

Usually, timing analysis means static timing analysis. In the context of this chapter also, timing analysis would mean static timing analysis, unless otherwise mentioned.

3.1 Static Timing Analysis

The concept of STA started around mid-1990s. Since then, static timing analysis (STA) has been gaining ground as the preferred method for timing analysis. STA is not dependent on the input vectors. STA involves analyzing the circuit topology and computing the time window within which various signals can reach various points in the circuit and then comparing it against the time when those signals are required at that point. As long as the range of time during which the signals arrive meets the required time, the design is clean – from STA perspective. The main reason why STA became popular vis-à-vis dynamic timing analysis is that it is

Fig. 3.1 A sample (abstract) circuit for STA basics

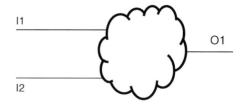

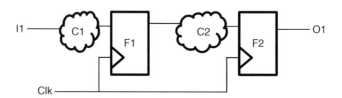

Fig. 3.2 Internals of the circuit shown in Fig. 3.1

much more exhaustive, as STA does not depend on input vectors' completeness. Starting with Chap. 5, we will learn the specifics of STA in much more detail. For the time being, let us consider a very abstract view of a circuit – as shown in Fig. 3.1.

STA would know at what times the signals can arrive at inputs *I1* and *I2* (and all other inputs). Now, STA would compute at what time these signals can reach the output *O1* (and all other outputs). STA would also know at what time the signal is required at the output *O1*. By comparing the time that the signal is available at *O1* – with the time when it is required – an STA tool will report whether the timing has been met or failed. If the circuit contains some flops, then, STA will also need to compute at what time each flop receives its clock and data. STA tool will then compare the data and clock arrival time on each flop with the setup and hold requirement for the flop. If the data arrival time does not meet the setup and the hold requirement for the flop, STA tool will report a violation. Let us dive slightly deeper into the circuit shown by Fig. 3.1. Let us say, Fig. 3.2 shows a small portion of the above circuit.

For this portion of the circuit, the data at *F1* should arrive before the setup window starts for *F1*, for each active edge of clock on *F1*. On the other hand, the data should arrive only after the previous data has been captured reliably (i.e., do not violate the hold time). This depends on the time at which triggering edge arrives on *F1*, the delay through the combinational circuit *C1*, and the input arrival time at *I1*.

Similarly, for the flop *F2*, the timing requirement will depend on the time at which triggering edge arrives on *F2*'s clock pin, the delay through the combinational circuit *C2*, as well as the time at which data starts from *F1* which in turn depends on the triggering edge at *F1*'s clock pin.

In Sect. 3.4, we will see how STA depends on delay calculation. Because STA depends so heavily on delay calculation, it can only be carried out after the gate-level netlist is available. There can be no STA at the RTL. Even for gate level, as the design progresses further into the flow, the estimates for routing delays and capacitive load become more accurate, which allows for better delay computation and so more accurate STA.

3.2 Role of Timing Constraints in STA

An STA tool gets the circuit description from the corresponding design description, HDL being the most commonly used form. It also takes in library inputs – mostly to know about technology-dependent characteristics, e.g., delay values through specific gates.

Another set of inputs that STA tools need are related to the arrival time and other characteristics of various signals at the inputs and the time at which various outputs are required. These inputs are provided through timing constraints. Timing constraints perform several roles during an STA.

3.2.1 Constraints as Statements

Certain constraints are simply a statement to the tool. The tool need not validate these statements for accuracy or correctness. The tool should simply use this as an input for itself, in order to validate that the timing would be met. Typically, this is information about some condition external to the design – something that the tool cannot determine on its own.

For the circuit given in Fig. 3.3, the following could be examples of constraints which are simply used as statements:

- The time at which input *I1* arrives at the boundary
- The time at which input *I2* arrives at the boundary
- The transition time (time needed for a signal to change its state from logic *0* to logic *1* or vice versa) for the inputs at *I1* and *I2*
- The load that has to be driven by the output *O1*

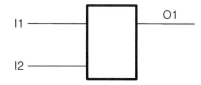

Fig. 3.3 Sample statement-type constraints

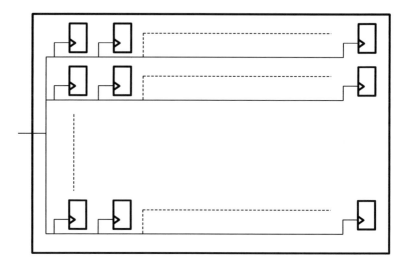

Fig. 3.4 Clock tree

3.2.2 Constraints as Assertions

On the other hand, there are certain constraints which act as assertions. The tool needs to validate that the design meets these constraints. This is what the STA is about. After doing various computations, if the tool finds that the timing meets these constraints, STA is considered as pass, or timing is considered to be met. Alternately, if these constraints are not met, STA or timing is considered as failed.

Looking at the same example circuit as shown in Fig. 3.3, an example assertion-type constraint would be:

- The time at which output $O1$ should be available at the boundary

3.2.3 Constraints as Directives

At some other times, constraints act as directives to certain tools. This happens for implementation tools, such as synthesis or place and route. These implementation tools take these constraints as a goal that they try to meet. Let us consider the circuit shown in Fig. 3.4.

For the example circuit, a clock tree synthesis tool will now put an elaborate network structure so that each of the 1,000s of flops can be driven by this network. An example directive for such a clock tree synthesis tool could be the delay for the clock network. The clock tree synthesis tool should implement the tree network in such a manner that the delay through the network meets the specified value.

3.2 Role of Timing Constraints in STA   21

Fig. 3.5 Output required time constraint

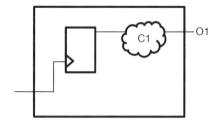

3.2.4 Constraints as Exceptions

There are certain constraints which do the opposite, namely, they specify scope for leniency. Suppose certain paths are constrained to meet certain timing. However, for some special reason, there is no need for those paths to meet those timings. Or, they can work well, even if they are given much more relaxed requirements. These are called timing exceptions. Chapters 11 and 12 have several examples of such timing exceptions.

3.2.5 Changing Role of Constraints

Sometimes, the same constraint changes its role. It could be a statement at one stage of the design, and at another stage of the design, it could become a directive.

Let us consider the circuit shown in Fig. 3.4 once again. There is a constraint which specifies the delay through the clock network. If somebody wants to do an STA before the clock network has been synthesized, then, this constraint (i.e., delay through the clock network) is simply a statement. The STA tool should simply assume that the delay through the clock network is the specified value.

Now, during clock tree synthesis, this same constraint becomes a directive. The clock tree synthesis tool needs to synthesize the clock network such that the delay through the clock network is equal to the value specified.

Once the clock tree synthesis has been done, the STA can compute the actual delay through the clock tree network. The same constraint becomes meaningless and should be thrown away!!!

Thus, the same constraint (delay through the clock network):

- Is a statement for pre-layout STA – where it assumes this as the delay through the network
- Is a directive for clock tree synthesis tool – which tries to realize the clock network such that its delay is within the specified range
- Is meaningless for post-layout STA

We can consider another example through Fig. 3.5.

For the given circuit, there will be a constraint specifying the output required time. For an implementation tool like synthesis, this constraint acts as a directive for implementing the combinational block *C1*. *C1* should be realized in such a manner that the delay through it still allows the signals to reach output *O1 at* the required time.

On the other hand, for an STA tool, this constraint acts as an assertion. The STA tool needs to check that the output is available at *O1 at* the required time. If the output is not available at the required time, it should report an STA or timing failure.

Thus, timing constraints could be an input statement or an assertion or a directive or even a relaxation (called timing exception). In most cases, the constraint itself does not indicate what role the constraint is playing. The role of the constraint has to be determined depending upon the context.

3.3 Common Issues in STA

The biggest issue with STA is a sense of security. There are many chips which were STA clean but did not operate correctly. A clean STA causes the designers to get too confident.

3.3.1 No Functionality Check

STA is only about timing analysis. It does not cover or verify anything else. It does no checking at all about functionality. A clean STA has no guarantee that the circuit will give the desired functionality. A clean STA would only mean that the circuit will operate at the specified frequency. Simulation, assertion-based checks, FPGA prototyping, etc. are several techniques that need to be used for ensuring correct functionality.

3.3.2 No Check on Statements

As already mentioned in Sect. 3.2, there are some constraints which specify designer intent and are simply treated as statements. The STA tools do not question these statements of designer intent. The STA tools simply assume these constraints to be true. In most cases, they might not even have the information required to verify these constraints. If a user makes an error in giving such constraints, the tool could give erroneous results. And, a design which does not meet desired timing might simply appear to be STA clean – due to user error in giving incorrect constraints. Even if there are portions of circuits where timing characteristics could be determined, user-provided constraints generally override what can be implicitly inferred. For example, for generated clocks (e.g., clock dividers), it might be possible to

3.3 Common Issues in STA

Fig. 3.6 Constraints accuracy requirements

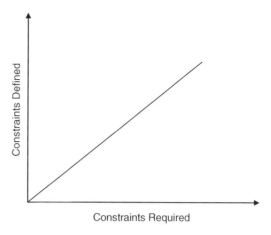

determine the clock characteristics. However, the user-specified characteristics override, even if they are incorrect. Details on such scenarios are explained in Chap. 6 on generated clocks.

3.3.3 Requirements to be Just Right

The constraints have to tread a very fine balance. Figure 3.6 explains this balance.

The horizontal axis denotes the constraints required for the desired frequency of operation. The vertical axis denotes the constraints actually applied on the design. Moving away from the origin signifies tighter constraints. The straight line at 45° denotes the ideal constraint. The constraints actually defined should be the same as what is actually required.

A point below the line means that the constraint applied is more lenient compared to what is required. This means the STA might appear to be clean, but it has been done against conditions more lenient than the actual requirements, and so the final device might not really work at the desired frequency.

An obvious solution is that the constraints should always be applied either on the line or above the line. This will ensure that the constraints are tighter than what is really needed. A clean STA here would ensure that the design is really timing clean. However, applying tighter constraints has its own set of problems.

- In order to meet stricter timing, the tools might insert higher-drive cells. These higher drive cells mean higher area and power.
- Sometimes in order to meet the tighter constraints, more efficient routing resources might be allotted to these paths, thus leaving out less efficient resources for the actual critical paths. Thus, the paths which really need better resources might get lesser priority, because another path had been given an unnecessarily tighter constraint.

- The worst implication of overly tight constraints is the timing closure problem. The design might not meet timing, because it has been specified a much tougher constraint than what is really needed. This might cause a lot of time to be spent unnecessarily – in trying to meet timing that is not even needed.

Thus, ideally the constraints should be applied just on the line – neither tighter nor lenient than the requirement. In case of a doubt, one has to err on the side of being a bit tighter.

3.3.4 Common Errors in Constraints

Some of the most common kind of mistakes that users might make while writing constraints include:

- Incorrect timing exceptions: Timing exceptions are the most misunderstood form of constraints. By specifying these exceptions, sometimes users provide leniency to paths, where it should not be provided.
- Incorrect clocks: The errors could be in clock period or waveform. However, the more common problem is around generated clocks. Most users concentrate only on the *divide_by* and *multiply_by* factors. However, the resulting generated clock waveform might have its positive and negative edges swapped – if not specified carefully.
- Not in sync with changing RTL: Many a time as a design keeps getting retargeted or reused, the RTL gets updated. However, the corresponding constraints are not updated. This might be because of lack of knowledge/awareness. But, more often than not, it happens because a user does not easily see the correlation of the RTL changes with the changes that are required in the constraints.

In subsequent chapters, we will see the implications of various switches and options on many of the commonly used constraints. Each of these as a corollary also indicates the kind of mistakes that users might make while writing the constraints.

3.3.5 Characteristics of Good Constraints

For any given timing requirement, there might be many ways in which constraints might be applied. A good set of constraints should satisfy the following conditions:

- First and foremost, these constraints must be obviously correct!! Section 3.3.3 has already explained the implication of incorrect constraints.
- The second most important characteristic is that by looking at the constraint, the intent should be clearly understood. This helps in review and catching mistakes, and it also insulates the constraints from minor changes in the design. For example, for the circuit in Fig. 3.7, one might declare a false path between *Reg1* and *Reg2*. Another user might declare a false path between *Clock1* and *Clock2*. And, yet

3.3 Common Issues in STA 25

Fig. 3.7 Intent of constraints

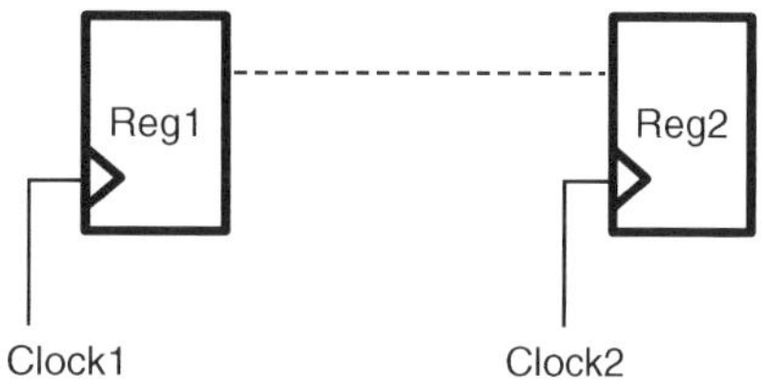

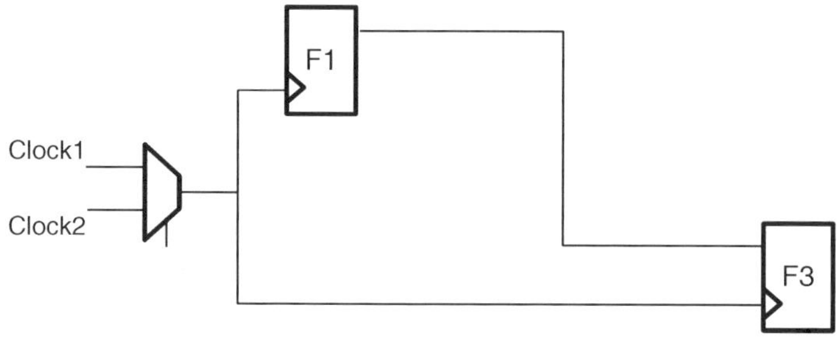

Fig. 3.8 Exceptions being invalidated due to RTL change

another user might declare the two clocks *Clock1* and *Clock2* in asynchronous groups. Though, from the STA perspective, all three have the same impact, but from understanding the intent perspective, the third one (asynchronous clock groups) is the best in conveying exactly the reason for the constraint (actually, an exception), while the first one (false path between the two registers) is the worst; it does not highlight that the false path is due to the clocks being asynchronous and has really nothing to do with the actual path between the two registers.

- Specifically for exceptions, care should be taken to write them in a manner such that potential minor changes in the RTL should not cause the exceptions to become invalid. Looking at Fig. 3.8, a mutually exclusive relationship between *Clock1* and *Clock2* seems reasonable. However, if the RTL is modified so that either of the clocks gets used before muxing also, then the above exception gets invalidated and is most likely to get missed being updated.
- The constraints should be written in a manner such that migrating to a new technology is the easiest. For example, instead of the input transition being specified in terms of a specific transition value, it is better to specify the driver cell. As technology changes, the new drive strength of the driver cell would be considered.
- Lastly, the constraints should be written in a manner that they are concise. It helps readability and reuse and most tools have lesser memory footprint for concisely written constraints. (Most SDC-based tools allow support for wildcards. However, an extensive usage of wildcard is not considered efficient, even though the constraints become concise. This is because wildcards could match additional objects also.)

Constraints written to satisfy the above characteristics are less error prone.

3.4 Delay Calculation Versus STA

Delay calculation and STA are two distinct and different things. STA depends heavily on delay calculation. Delay calculation has applications other than STA also. Delay calculation (for a given process, temperature, and voltage conditions) performs the following activities:

- Compute the delay through specific gates, nets, net segments, paths, etc.
- Compute the slew at the output of specific gates.
- Compute the slew degradation as the signal passes through a wire (which in turn becomes the slew at the input pin of the next gate).

After this, the delay values might be used directly by an STA tool for its analysis. In this case, the delay calculator is a part of the STA tool itself. Or, the various delay values might be written out in an *SDF* (standard delay format) file. This SDF file is read back into the STA tool. In this case, the delay calculation and STA are performed by two different tools.

An STA tool's functionality can be mentioned in an oversimplified manner to be as follows:

1. Collects all statement-type constraints, e.g., when are the inputs available
2. Passes some statement-type constraints to delay calculators, e.g., input transition time
3. Looks at the circuit topology to identify various timing paths
4. Obtains the path delays from the delay calculator
5. Combines 1 and 4 to compute the arrival time of the signals at the desired points
6. Compares the signal arrival time with assertion-type constraints which tell when should the signal be available there
7. Provides a result – based on actual arrival time and the required time

Clearly, the correctness of the STA depends very heavily upon the underlying delay calculation.

3.5 Timing Paths

A timing path means a path through which a signal can continue to traverse, without having to wait for any other triggering condition. Along a timing path, the signal only encounters the delay through the circuit elements.

3.5.1 Start and End Points

The point at which a signal's timing starts is called a start point. Thus, for a given circuit, all inputs act as start points. The point at which a signal has to be timed is called an end point. Thus, all outputs act as end points.

3.5 Timing Paths

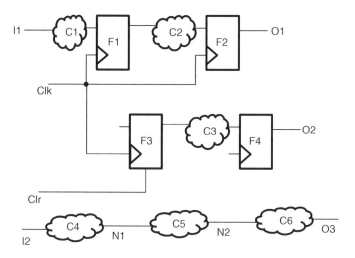

Fig. 3.9 Timing paths

At the registers, the *D* input has to wait for the *clock* trigger to arrive. So the transitions which reach till *D* now have to wait. The timing for the signal propagating further will now depend on when the *clock* arrives. This is the place where the *D* should be checked for meeting setup and hold requirements. Thus, the timing path ends here. So registers also act as end points.

Similarly, registers also act as start points. A signal will start from the *Q* pin of the register and then propagate forward. Strictly speaking, the timing starts from the *clock* source, reaches the *clock* pin of the flop, and then goes to the *Q* pin of the flop and then proceeds further. So, in a strict sense, the register is not really a start point. However, for most practical purposes, registers are referred as start points. Though during actual analysis, the path tracing starts from the clock source, for the circuit shown in Fig. 3.9.

Start points are:

- Primary inputs (*I1, Clk, Clr, I2*)
- Registers (*F1, F2, F3, F4*) – actually the clock sources of these registers!!

End points are:

- Primary outputs (*O1, O2, O3*) – this is where the check has to be made that the signals are available at these points at the desired time
- Registers (*F1, F2, F3, F4*) – this is where the *D* input (and any other synchronous inputs) has to be checked for meeting setup and hold requirements

Let us look more closely at flop *F3*. A signal reaching the asynchronous *clear* pin of the flop need not wait here for any other trigger. It can simply continue through the *Q* pin of the flop and onwards. Thus, timing path need not end here. So while *D* pin of a flop acts as an end point, an asynchronous *clear* or *set* pin might not be an end point.

3.5.2 Path Breaking

A user can insert additional start and end points anywhere in the design by specifying some starting conditions or a checking condition. By specifying a starting point or a checking point, the path gets broken at that point.

For example, in the same Fig. 3.9, let us say, a user has specified that the maximum delay from *N1* to *N2* should be some value. In such a case, a check has to be made at *N2*. Thus, *N2* becomes an additional end point. And it also becomes a start point for the next segment of the path, namely, *N2* to *O3*. Similarly, since the signal tracing is starting from *N1*, it becomes a start point. And it also becomes an end point for the previous segment of the path, namely, *I2* to *N1*.

Now, if a user had specified a delay for *I2* to *N1*, then, *N1* would become an end point. It would also become a start point for the next path segment. However, *I2* remains only a start point. It does not become an end point, because there is no previous path segment.

Looking again at flop *F3*, a timing path would mean *Clr* → *F3's clear* pin → *F3's Q* pin → *C3* → *F4's D* pin. However, if a recovery – removal check is applied on the *clear* pin of *F3*, then the path gets broken at that point.

3.5.3 Functional Versus Timing Paths

The path topology along which a transition travels from the start point till the end point is called a timing path. It is different from a functional path. A functional path means the topology along which a signal travels.

For example, for the circuit in Fig. 3.9, a signal starting from I1 will traverse through *C1* and then reach the *D* pin of *F1*. After that, when the conditions are right (viz., a *clock* trigger), it will cross *F1* and continue on to *C2* and onwards. So this is a functional path.

However, this is not a timing path. Because, irrespective of when the signal arrives at *D*, we do not know when will it proceed ahead to cross the flop. It would depend on when the clock trigger arrives at the flop. Timing path means, when a signal arrives, we can say when the next transition in the path will happen. For example, when the clock trigger arrives, we know the next transition will happen at the flop output.

The actual transition may or may not happen. For example, if the *D* has the same value, the *Q* would remain unchanged. The timing path indicates the possibility of a transition.

Thus, timing paths indicate the sequence along which transition would follow without having to wait for any other event to happen. The actual value could be coming from somewhere else. Functional paths indicate the sequence in which values would change. The timing of the value change might still be controlled by something else. Sometimes, a timing path and a functional path could overlap, e.g., paths through a combinational circuit.

3.5.4 Clock and Data Paths

A path which feeds into a data pin of a flop is called the data path. So data path could be:

- From a flop output to another flop's data input
- From a primary port to a flop's data pin
- From a flop output to an output port (to be fed into a flop outside this design)

A path which feeds into the clock pin of a flop is called the clock path. So clock path could be from the clock port till the flop's clock terminal, from the output of a clock divider or clock generator circuitry till a flop's clock terminal, etc.

We have used the word "flop" to mean any synchronous element, e.g., memory.

3.6 Setup and Hold

STA is mostly about setup and hold analysis. In general terminology, setup means the time before a clock edge before which the data should be stable on the *D* (or any other synchronous) pin of a flop. Hold means the time after a clock edge for which the data should be held stable.

3.6.1 Setup Analysis

In STA world, setup means checking that the latest data is available before the required time. Thus, setup check can be made at any end point – not just the flop. A setup check would be made even at outputs.

Setup can be defined in a more generic way as follows: The data needs to be set up and available before some reference event. In the case of a flop, the reference event is the clock trigger. In the case of other end points, the reference event is the "time at which the data is expected to be available at that point."

Let us consider the circuit shown in Fig. 3.10.

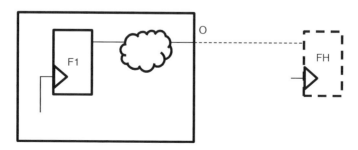

Fig. 3.10 Setup check at an output

For the output *O*, let us assume that the output is required to be available at *6* time units after the clock edge. Assume that the clock has a period of *10*. This requirement can be thought of as a setup requirement of *4* (*10 – 6*) on a hypothetical flop (*FH* – shown in dotted lines), which sits beyond the output port *O*. So this setup check can be made at the output.

3.6.2 Hold Analysis

Similarly, hold means checking that the earliest time a new data can disturb the current signal after the stability requirements are met for the current signal. Thus, like setup, hold check can be made at any end point, including at outputs – not just the flop.

Hold can be defined in a more generic way as follows: The data needs to be held and left undisturbed after some reference event. In the case of a flop, the reference event is the clock trigger. In the case of other end points, the reference event is the "time at which the data is allowed to be changed at that point."

Once again let us consider the circuit shown in Fig. 3.10. For the output *O*, let us assume that the output is required to be held for 2 time units after the clock edge. This requirement can be thought of as a hold requirement of *2* on the hypothetical flop (*FH*), which sits beyond the output port *O*. So a hold check can be made at the output.

3.6.3 Other Analysis

Setup check ensures that the slowest moving data also reaches and meets the criterion of being setup in time. Thus, for the data path, it computes the maximum delay. Thus, it is also called *max analysis*. Since setup check considers the latest arriving data, it is also called *late analysis*.

Similarly, hold check ensures that even the fastest moving data should not disturb the data, while it is expected to remain stable. Hence, for the data path, it computes the minimum delay. Thus, it is also called *min analysis*. Since hold check considers the earliest arriving data, it is also called *early analysis*.

Setup – hold analysis of STA is also called *min – max analysis* or *early – late analysis*. It is more important to be familiar with the concepts, rather than getting too worried about the terminology. Sometimes, different tools might use different terminology for the same concept. Or, sometimes, even the same terminology is used to refer to different concepts in different contexts.

Besides setup or hold analysis, STA performs pulse-width, recovery, removal analysis, etc. also.

3.7 Slack

Slack refers to any additional margin over and above the requirement. Say, a signal is required to be available before time 6 (setup analysis). The last signal arrives there at time 4. So the signal has a margin to take an additional 2 time units, without risking the operation of the design. This 2 is called *setup slack*.

Similarly, say a signal is required to be kept stable till time 2 (hold analysis). The earliest a new signal reaches there is at time 5. So the signal has a scope for getting faster by another 3 time units. This 3 is called *hold slack*.

Setup slack = data setup requirement – last arriving signal
Hold slack = earliest arriving signal – data stability requirement

Figure 3.11 explains this better.

Let *E1* be the edge, where data needs to be captured. *S* represents a time which is ahead of *E1* by a duration equal to setup requirement. So the latest data should reach the end point before time *S*. Let us say that the last arriving change happens at time *Ma*. So the setup slack is the duration *Ma – S* (measured by *S – Ma*).

Just like the data has to arrive before *S*, similarly, the data should not arrive so early that it can interfere with the capture of the data at the previous edge. Let *E0* be the previous edge, where previous data is supposed to be captured. For a reliable capture of the previous data, the current data should not disturb it till the hold requirement from *E0*. Let *H* represent a time, which is hold time after *E0*. So the current data can come only after *H*. Let us say that the earliest arriving change happens at time *Mi*. So the hold slack is the duration *H – Mi* (measured by *Mi – H*)

A positive slack means timing has been met. And, a negative slack means the timing has not been met. Depending upon the tool being used, the exact format of

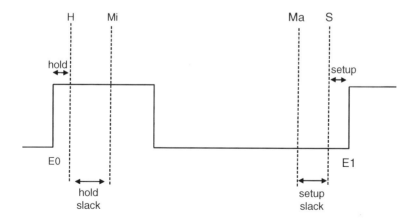

Fig. 3.11 Slack

the report would be different. However, for any of the tools, they would trace the delay (maximum for setup analysis and minimum for hold analysis) through the path and also trace the time required for the signal. It would then compare the two numbers and provide the slack number (either positive or negative). The path delay or the required time computation also considers the delay along the path of the clock that is used to trigger the launch of the data at the start point and the capture of the data at the end point.

3.8 On-Chip Variation

Consider the circuit shown in Fig. 3.12.

We want to do a setup analysis at flop *F2*. So the slowest data path has to be considered. So we will consider the maximum delay for:

- $Clk \rightarrow Q$ delay of the flop *F1*
- Combo logic *C1*
- Interconnect net

However, the data moves across the flop only when the clock trigger arrives. Thus, we have to consider the clock path till flop *F1* also with maximum delay. This same clock is also reaching flop *F2* and acts as the reference event for the setup check. If we treat the clock path with maximum delay, then the reference event also gets delayed, which might allow additional slack for the data to arrive. So, for the clock path for the flop *F2*, the minimum delay is considered.

Effectively, for the same clock, one segment ($A \rightarrow B \rightarrow F1$'s *CLK* pin) has been considered as slowest path, while the other segment ($A \rightarrow C \rightarrow D \rightarrow F2$'s *CLK* pin) is considered as fastest path.

This differential treatment for the different segments of the same network accounts for any variation on different portions of the same chip and is called on-chip variation.

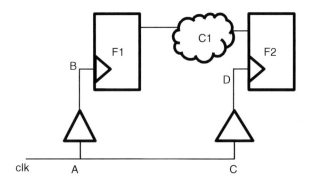

Fig. 3.12 On-chip variation

For the hold analysis, the data path has to be the fastest, so the segment of the clock network which feeds into *F1* will also be considered at fastest speed, and the segment which feeds into *F2's CLK* pin will be considered at slowest speed.

This on-chip variation reduces the slack. There is a segment of the clock network which is common to both the launch and the capture flops (from the *Clk* port till *A*). The clock traversal through this segment will have the same delay whether for capture flop or the launch flop. This segment which is common to both the flops is considered to have same delay values. This prevents over-pessimism. Some tools apply on-chip variation for the whole clock path, including the common segment. After that, they apply a correction factor to compensate for the differential delay considered in the common segment. This gets reflected in a term called *clock network pessimism reduction* or *clock tree pessimism reduction*.

3.9 Conclusion

Constraints provide a way for the user to specify their timing intent to an STA tool as well as implementation tools. STA is used to do setup and hold analysis and compute the slack for the paths being analyzed. In order to do this analysis, STA depends very heavily on delay calculations.

Chapter 4
SDC Extensions Through Tcl

The term "Synopsys Design Constraints" (aka SDC) is used to describe design requirements for timing, power, and area and is the most commonly used format by EDA tools used for synthesis, static timing analysis, and place and route. This chapter provides a brief history of timing constraints and an overview of the SDC format.

4.1 History of Timing Constraints

Timing constraints were introduced in the early 1990s. These were mainly used for specifying design characteristics that could not be captured in the HDL and were used to drive synthesis. At that time these were commands for Design Compiler® to provide guidance to synthesis tool to optimize the design on area versus performance curve. When PrimeTime® incorporated the concept of Tcl, these constraints were modified to be an extension of the Tcl format. This set was called SDC or Synopsys Design Constraints. Over the years this set was extended to capture design requirements for power as well.

SDC commands are based on the Tcl language. "Tool Command Language" (aka Tcl) is a very popular scripting language commonly used in developing applications for user interface and embedded systems platform. By making SDC an extension to Tcl, tool-specific commands can be intermixed with native Tcl constructs like variable, expressions, statements, and subroutines, making it a very powerful language for implementation tools. Today most implementation and STA tools use SDC as the standard format for capturing design intent for area, power, and performance. However, specific implementation tools may also use additional commands outside this set in order to guide their specific algorithms or capabilities. These commands are generally referred to as the non-SDC commands.

For example, Design Compiler and PrimeTime have a number of non-SDC commands that are used in conjunction with SDC constructs. Some of these extensions facilitate the analysis of constraints, e.g., *collections*, which provide a way to iterate over a list of design objects. While some non-SDC commands are still used to capture design intent, they are periodically included in SDC revisions based on its usefulness and popularity. For example, *set_clock_groups* which was a concise way of representing domain relationship between clocks was a non-SDC command until SDC standard *1.7* (March 2007).

Today Synopsys provides SDC as an open source format for describing timing intent. However, changes to the format are still controlled by Synopsys. SDC documentation and parsers can be downloaded for free from http://www.synopsys.com/community/interoperability/pages/tapinsdc.aspx. At the time this book was being written, the latest SDC version was SDC 1.9 which was introduced in Dec 2010.

4.2 Tcl Basics

Most EDA tools today support Tcl as their command shell. This shell along with its SDC extension facilitates users to create Tcl-based wrappers in order to drive various queries and instructions to the tool. So it is good to have a reasonable understanding of Tcl. A basic explanation of Tcl is given here. For a detailed understanding of Tcl, the reader is advised to read a book on Tcl.

Tcl is a commonly used scripting language that was developed in 1988 by John K. Ousterhout from University of California, Berkeley. Unlike a compiled language, where the language is parsed and compiled into machine code before execution, Tcl is an interpretive language where each statement is parsed sequentially and executed right away. Hence, the language stops at the first error it encounters in a script.

Tcl follows some basic semantics in its scripting. Each Tcl statement ends with a newline character or semicolon. If a statement spans multiple lines, then to continue on the next line, a backslash is provided at the end of the line. Every statement and its arguments are treated as strings. A string with more than one word enclosed in double quotes or braces is considered a single unit. A statement beginning with # is considered a Tcl comment. Here are a few examples:

puts "Hello, World!"; # This is a comment
puts {Hello, World!}

Each of these statements will print "Hello, World!" on stdout. Words separated by white space are treated as multiple arguments to the statement.

puts Hello World!

This will give an error indicating it cannot process the arguments. Error will show up as:

cannot find a channel named 'Hello'

4.2 Tcl Basics 37

Tcl has many kinds of language constructs. These are:

- Variables
- List
- Expressions and operators
- Control flow statements
- Procedures

4.2.1 Tcl Variables

Tcl variables are a string of ASCII characters. Numbers are represented as ASCII characters as well. Variables are assigned using the *set* command. For example,

set abc "1234"; # Here set is the command, abc is the variable and 1234 is the value assigned to it.

If you need to evaluate a variable, you need to use the *dollar ($)* symbol. For example, puts $abc will print 1234.

A variable can be treated as an array, if an index is used along with variable name. An index doesn't have to be an integer; it can also be a string. For example,

set def(1) 4567; # Here def is the array variable and the index is set to 1.
set def(test) 5678; # Here the index is test

This index *"1"* of array is set to value *4567*. Note the value *4567* doesn't appear in double quotes. In Tcl *"4567"* and *4567* both mean the same since everything is treated as a string. To get values of indices used in an array, use the command *array names*. For example, *array names def* will return the values *1* and *test*. To evaluate an array variable, you can use the same mechanism, but the index needs to be specified. Let us consider the examples below:

puts $def(1); # Valid Command
puts $def(test); # Valid Command
puts $def; # Invalid command

Since $ has a special meaning in evaluating a variable, to actually print this symbol, it has to be preceded by a backslash (\). For example, *puts "I have a $bill"* will result in error if variable *bill* is not defined or will simply print the value of the variable. The correct way to use this would be: *puts "I have a \$bill"*. The evaluation of a variable is also disabled within a brace. *puts {I have a $bill}* will not try to evaluate the variable *$bill*. However, there are exceptions to this rule as illustrated by an example in Sect. 4.2.3.

If you are trying to set a variable from another command, then enclose the command in square braces. Anything in square braces is evaluated before it is used. However, square braces within braces are not evaluated.

set x [set y 100]; #This will set the value of x to 100
set x {[set y 100]}; #This will set the value of x to [set y 100]

4.2.2 Tcl Lists

Lists in Tcl are a collection of objects. Like any list, you can add to a list, index into a list, and search within a list. Here are few examples:

 # The following creates a list
set gates [list AND OR NOT NAND NOR]
set gates {{AND} {OR} {NOT} {NAND} {NOR}}
set gates [split "AND.OR.NOT.NAND.NOR" "."]

 # To add another item to a list, use lappend
set gates [lappend gates XOR]

 # To search within a list, use lsearch. This returns the matching indices of the list
 # Returns 2, which is the index in the list
puts [lsearch $gates NOT];

 # Returns -1, since there is no match found
puts [lsearch $gates XNOR];

4.2.3 Tcl Expression and Operators

Expressions in Tcl are evaluated using the *expr* command. Let us consider the following example:

set x 10;
 # Both statement will return the value of 30
expr $x + 20
expr {$x + 20}

In the example above, both of the *expr* commands will result in the same value. However, Tcl recommends the expression with braces as it facilitates faster execution.

Since expression evaluation is very closely related to operator used, Tcl language provides a comprehensive support for logical and arithmetic operators. Table 4.1 shows the list of operators supported.

4.2.4 *Tcl Control Flow Statements*

Tcl control flow statement consists of the following category of constructs:

- Iterating over lists
- Decision making
- Loops
- Subroutines

4.2 Tcl Basics

Table 4.1 List of supported operators in Tcl

Operators	Description
- + ~ !	Unary minus, unary plus, bit-wise NOT, logical NOT
+ - * /	Addition, subtraction, multiplication, division
**	Exponents
< > <= >= == !=	Relational operators: less, greater, less than or equal, greater than or equal, equality, and no equality
eq, ne	Compare two strings for equality (eq) or inequality (ne)
in, ni	Operators for checking if a string is contained in a list (in) or not (ni). Returns 1 for true and 0 for false
& | ^	Bit-wise AND, OR, XOR
&&, ||	Logical AND, OR
<< >>	Left or right shift

4.2.4.1 Iterating over Lists

To iterate over lists, Tcl provides the *foreach* construct.

set gates [list AND OR NOT NAND NOR XOR]
set index 1
foreach element $gates {
puts "Gate $index in the list is $element"
incr index; #This increments the index
}

This will generate the output as:

Gate 1 in the list is AND
Gate 2 in the list is OR
Gate 3 in the list is NOT
Gate 4 in the list is NAND
Gate 5 in the list is NOR
Gate 6 in the list is XOR

The Tcl *foreach* gives the user a unique ability to iterate over multiple lists simultaneously. Let us consider the example below:

set allgates {}
foreach gatelist1 {AND OR XOR} gatelist2 {NAND NOR XNOR} {
lappend allgates $gatelist2 $gatelist1
}
puts $allgates

This will store the following value in variable *allgates* "NAND AND NOR OR XNOR XOR". As you can see, this iterator gives you the ability to mix the items from different lists.

4.2.4.2 Decision Making

Tcl provides *if-elseif-else* construct to provide decision-making ability. Let us consider the following examples:

if { $frequency < 330 } {
puts "Chip will function well, but slower than expected"
} elseif { $frequency > 330 } {
puts "Chip will not function"
} else { puts "Chip will function optimally" }

It is important to use the braces {} carefully. Each segment has its own pair of braces.

4.2.4.3 Tcl Loops

Tcl provides *for* and *while* statements when a program wants to loop and terminate on a condition. It also provides two additional constructs *break* and *continue*. *break* is used to terminate a loop prematurely, while *continue* is used to stop the execution of the code for the current loop and reevaluate the condition of the loop. Please note *break* and *continue* can also be used with *foreach* while iterating over lists. *for* and *while* follow the semantics as shown below:

for <initial value> <condition> <next step> {
<statement body>
}
while <condition> {
<statement body>
}

4.2.4.4 Tcl Procedures

Tcl procedures are written using *procs*. The value is returned from the procedure using a *return* statement. Let us consider the example below:

proc sum {addend1 addend2} {
set value [expr {$addend1 + $addend2}]
return $value
}
 # Calling the procedure
set x [sum 5 10]
puts $x # Will print 15

4.3 SDC Overview

Table 4.2 Commonly use Tcl commands

Command	Description
open/close	File handle to open and close files
	# open a file "file.txt" in write mode
	set fhandle [open "file.txt" w]
	# close file handle
	close $fhandle
gets/puts	Get a string or prints a strings from/to stdin/stdout. When used with a file handle as defined above, it will fetch from the file or write to a file
catch	Command to capture error from a command, so as to prevent the Tcl shell from aborting
	set file_name "file.txt"
	if { [catch {open $file_name w}] $fid} {
	# Error from open command is in $fid
	puts stderr "Error: $fid \n"
	exit 1
	}
info	Command used from Tcl interpreter to get information
	info commands <pattern> : Returns a list of the commands, both internal commands and procedures, whose names match pattern
	info exists <name> : Returns 1 if name exists as a variable, otherwise returns 0
	info procs <pattern> : Returns a list of the Tcl procedures that match pattern
source	Command to source a Tcl file or script
incr	Command to increment an index
exit	Command to exit from Tcl shell

Procedures can also be defined to define or set default values for arguments as *proc sum {{addend1 10} {addend2 20}}*.

4.2.5 Miscellaneous Tcl Commands

Table 4.2 shows some of the frequently used Tcl commands.

4.3 SDC Overview

The constraints in SDC format can be broadly categorized as:

1. Constraints for timing
2. Constraints for area and power
3. Constraints for design rules
4. Constraints for interfaces

Table 4.3 Constraints for timing

create_clock	create_generated_clock	set_clock_groups
set_clock_latency	set_clock_transition	set_clock_uncertainty
set_clock_sense	set_propagated_clock	set_input_delay
set_output_delay	set_clock_gating_check	set_ideal_latency
set_ideal_network	set_ideal_transistion	set_max_time_borrow
set_resistance	set_timing_derate	set_data_check
group_path	set_drive	set_load
set_input_transition	set_fanout_load	

5. Constraints for specific modes and configurations
6. Exceptions to design requirements
7. Miscellaneous commands

Some of the Constraints can fall in more than one category.

4.3.1 *Constraints for Timing*

Constraints for timing provide guidance on design parameters that affect operational frequency. It includes commands to specify clock characteristics, delays on port, and pins and paths. Table 4.3 shows the list of constraints in this category.

4.3.2 *Constraints for Area and Power*

Constraints for area and power include commands that provide guidance on the area a design must fit within and power requirements for optimization. Table 4.4 shows the list of constraints in this category.

4.3.3 *Constraints for Design Rules*

Constraints for design rules include commands that provide guidance on some of the requirements of the target technology. Table 4.5 shows the list of constraints in this category.

4.3 SDC Overview

Table 4.4 Constraints for area and power

set_max_area	create_voltage_area
set_level_shifter_threshold	set_max_dynamic_power
set_level_shifter_strategy	set_max_leakage_power

Table 4.5 Constraints for design rules

set_max_capacitance	set_min_capacitance
set_max_transition	set_max_fanout

Table 4.6 Constraints for interfaces

set_drive	set_driving_cell	set_input_transition
set_load	set_fanout_load	set_port_fanout_number
set_input_delay	set_output_delay	

4.3.4 *Constraints for Interfaces*

Constraints for interfaces include commands that provide guidance on the assumptions design needs to make about blocks it will be connected to or interacting with in a subsystem or chip or SoC. Table 4.6 shows the list of constraints in this category.

4.3.5 *Constraints for Specific Modes and Configurations*

Constraints for what-if analysis include commands that help designers make assumptions on the value allowed on ports and pins which facilitate better optimization of the design for a specific mode by specifically ruling out conditions that won't be possible in a specific mode of operation or in any of the modes. Table 4.7 shows the list of constraints in this category.

4.3.6 *Exceptions to Design Constraints*

This category includes commands that help designer relax the requirements set forth by other commands thereby providing scope for leniency. Table 4.8 shows the list of constraints in this category. The commands in the table marked with asterisk can also be used to provide additional tightening (rather than relaxing).

Table 4.7 Constraints for specific modes and configurations

set_case_analysis	set_logic_dc
set_logic_zero	set_logic_one

Table 4.8 Exceptions to design requirements

set_false_path	set_multi_cycle_path	set_disable_timing
set_max_delay*	set_min_delay*	

Table 4.9 Miscellaneous commands

set_wire_load_model	set_wire_load_mode
set_wire_load_selection_group	set_wire_load_min_block_size
set_units	set_operating_conditions
sdc_version	

4.3.7 *Miscellaneous Commands*

The rest of the SDC commands fall in this category. These commands provide guidance on operating conditions, wire load model, units, and version of the timing constraints. Table 4.9 shows the list of miscellaneous commands.

Details on the use model of these constraints and their application will be explained in the later chapters.

4.4 Design Query in SDC

In addition to these aforementioned categories, SDC standard also provides a way to access information about the design to facilitate design query, traversal, and exploration. These are used in conjunction with SDC constraints and Tcl scripting to effectively apply the design requirements at the right place in the design. Table 4.10 provides a brief description of commands in this category.

4.5 SDC as a Standard

As a standard, SDC is very loose. It pretty much mentions the commands and their arguments/switches. For the arguments and switches also, the standard only says which of these are optional and which are mandatory and what could be the types for the values being specified.

4.5 SDC as a Standard

Table 4.10 Commands for design query

Command	Description
get_cells	Returns the instance of the design or library cell
get_ports	Returns the input, inout, and output ports of the design
get_pins	Returns the instance of a port of design or library cell pin
get_nets	Returns net connected to the port or pin
get_clocks	Returns clocks in the design
all_inputs	Returns all input and inout (which are theoretically inputs also) ports in the design
all_outputs	Returns all output and inout (which are theoretically outputs also) ports in the design
all_registers	Returns all registers in the design
all_clocks	Returns all clocks in the design
get_libs	Returns the list of libraries
get_lib_cells	Returns the list of cells in the library
get_lib_pins	Returns the list of pins on the library cell
current_design	Sets the scope of the design for the subsequent commands and queries Report the current scope, if no argument is given

The standard itself does not say anything on how to interpret any of these commands or the switches or the values. It does not say anything as to which combinations are legal and which combinations are not. It does not specify if an optional switch is not applied what the default is. However, most tools interpret most of these commands and their options and arguments in a consistent manner. The behavior being same has been dictated more by the user community, rather than by the standard itself. In the following chapters, for many commands and their switches, their behavior has been described. This is the behavior as exhibited in the tools. The standard may not necessarily specify that behavior. However, considering that all the tools pretty much exhibit the same behavior, for all practical purposes, it might as well be considered as the behavior dictated by the standard.

Considering that the standard does not enforce any specific treatment/interpretation, occasionally, some tools could exhibit some difference in behavior, especially around those commands or options which are not used very commonly. The authors recommend users to write their SDC using more commonly used constructs – which ensure interoperability across tools from different suppliers. The authors also recommend specifying all requirements explicitly, rather than depending on tool defaults, since the defaults are also not specified by the standard.

For the sake of simplicity, in most examples in this book, the time unit has been assumed to be *ns* (nanoseconds). That is, for explaining the given values, we have simply mentioned *ns*, rather than specifying the units for each set of commands. For your SDC, the actual units would be as specified in your library, which could be different from *ns*. Treatment of units is also mentioned in Chap. 16 of the book.

4.6 Conclusion

Since SDC is Tcl compliant, most tools that support SDC support native Tcl. A user could combine the power of Tcl programming with SDC to specify design requirements in a very effective manner. Today SDC is the industry standard format that helps drive implementation tools to meet a design's timing, power, and area requirements.

Now that we have some understanding of Tcl and how constraints drive various tools, from next chapter, we will learn how to write the actual constraints.

Chapter 5
Clocks

A synchronous design is one where a control signal triggers the circuit to transition from one state to another. Such a trigger can happen at the positive or negative edge or both edges of the control signal. At the appropriate trigger edge, which may be active high or active low, input, outputs, internal registers, and nodes reach a stable state. Such a control signal which acts as a trigger for a synchronous design is called a *clock* and the edge on which the design triggers is called the *active edge* of the clock. A circuit that generates such a clock signal is called a clock generator.

A clock has a specific periodicity in its behavior which controls the timing within the design and is identified by its characteristics and the way it is used. These characteristics are:

1. Period
2. Active edge
3. Duty cycle

The other characteristics applicable to any signal like edge rate, rise and fall times are applicable to clocks as well.

5.1 Clock Period and Frequency

The period of a clock indicates time after which a clock will repeat its behavior. Let us consider the waveform of a signal shown in Fig. 5.1.

At time $t=5$, the signal goes from state *zero* (0) to state *one* (1). It remains in that state for the next *5ns*. At $t=10$, it goes back to state 0 and stays that way until $t=15$. At time $t=15$, this pattern starts repeating. Such a signal is said to have a *period* of *10ns* and can be used a clock. Period is typically represented in units of time, which could be seconds (abbreviated as s) or its derivatives like nanosecond (ns = 10^{-9} s) or picoseconds (ps = 10^{-12} s).

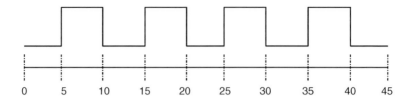

Fig. 5.1 Clock waveform

The rate at which a clock's active edges appear is called the *frequency*. If the period of clock is T, then its frequency is defined as $1/T$. This is represented in units of Hertz (abbreviated as Hz) or its derivatives like megahertz (MHz = 10^6 Hz) or gigahertz (GHz = 10^9 Hz). The speed/performance of circuit is often confused with the clock period. A small clock period in a device indicates higher clock rate or frequency; implying that the design has higher number (within a given duration) of active edges of a clock; and thus, on a given active edge, it needs to move to the next stable state much faster, before the next edge triggers. However it should be noted that clock frequency alone cannot be considered a measure of a device's performance. In today's world we are used to seeing the clock rate of microprocessors in MHz and GHz. Clock frequency is one of the parameters that effects a design's performance, but higher values of these parameters alone may not necessarily indicate that microprocessor will run faster and be more efficient. There are other indicators like the architecture and pipelining in a design, which could impact the performance, power consumption, etc. This misconception is generally referred to as the "megahertz myth," a term coined by the late Steve Jobs of Apple in 2001 when he compared the performance of 867 MHz PowerPC processor to 1.7 GHz Pentium processor.

5.2 Clock Edge and Duty Cycle

A clock will have positive and negative edge. Positive edge is when a clock transitions from state *0* to state *1*. Let us consider the waveform of the clock as shown in Fig. 5.2. The clock has a positive edge at time = *{0, 10, 20, 30 ....}*. Negative edge is when a clock transitions from state *1* to state *0*. In Fig. 5.2, the clock has a negative edge at time = *{7, 17, 27 ...}*.

A clock is said to be in *high transition* as it changes state from *0* to *1* and is said to be in *positive phase*, while it holds this value until next change. The clock shown in Fig. 5.2 is in positive phase from $t=0$ to $t=7ns$. When a clock changes value from *1* to *0*, it is said to be in *low transition* and in *negative phase*, while it holds this value until the next change. For the circuit in Fig. 5.3, the positive edge of the clock at time $\{t=0, t=10 ...\}$ will trigger the circuit. However for the circuit in Fig. 5.4, the negative edge of the clock at time $\{t=7, t=17 ...\}$ will trigger the circuit. *Duty cycle* is defined as the percentage of the time clock spends in positive phase as a fraction of its total time period. Therefore for the positive edge-triggered design, the

5.2 Clock Edge and Duty Cycle

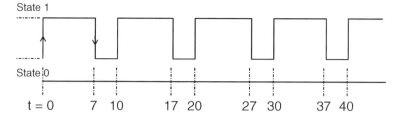

Fig. 5.2 Clock waveform with uneven duty cycle

Fig. 5.3 Positive edge-triggered circuit

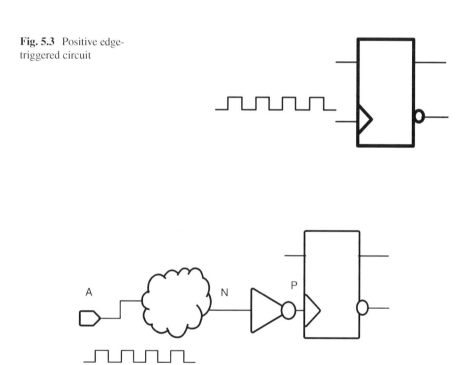

Fig. 5.4 Negative edge-triggered circuit

clock has a duty cycle of 70 %, since it spends 70 % of its time (7 out of the *10ns*) in its positive phase. The amount of time also has direct implication on how long a circuit is ON, thereby affecting its power consumption. The difference in the time interval between the positive and negative edge of a clock is referred to as the high pulse width, which is same as the time the design is in high state. This is also applicable to latch-based designs.

5.3 create_clock

The SDC command for specifying clocks in a design is *create_clock*. The BNF grammar for the command is:

create_clock -*period* period_value
  [source_objects]
  [-*name* clock_name]
  [-*waveform* edge_list]
  [-*add*]
  [-*comment* comment_string]

5.3.1 Specifying Clock Period

-*period* option is used to specify the period of the clock. The unit of clock period is inferred from the library time units. In all the examples in this book, time unit has been assumed to be *ns*. Period must have value greater than zero.

The *set_units* command mentions the units used in the SDC file. Details of this command are explained in Chap. 16.

5.3.2 Identifying the Clock Source

create_clock is generally specified on design objects which are used as sources of clock. These source objects can be port, pin, or net. For example, in Fig. 5.4, the source can be port *A*, net *N*, or pin *P* of the flip-flop. When defining a clock on a net, ensure that net has a driver (either a pin or a port). Otherwise the clock will not have a source. A clock can potentially have more than one source. This is mainly used when design has to support clock switchover for redundancy or different mode of operation. Clock switch over is a feature generally available in PLLs where in the redundant clock can turn on, if the primary clocks stop running.

Let us consider the Fig. 5.4 and let us assume that a clock with *10ns* period is driving the circuit. This can be represented as:

 # Represents the port as clock source
create_clock -period 10 [get_ports A]

OR

 # Represents the net as clock source
create_clock -period 10 [get_nets N]

OR

 # Represents the pin as clock source.
 # Assuming, flop instance name is FF
create_clock -period 10 [get_pins FF/P]

5.3 create_clock

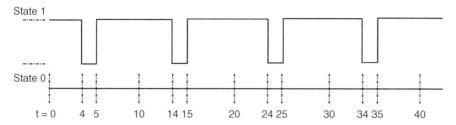

Fig. 5.5 Clock waveform with low pulse

5.3.3 Naming the Clock

Each clock definition is given a name. This is specified as string using the -*name* option. When -*name* is not specified explicitly, the tool might specify a name on its own – typically the object on which the clock is declared. For the first example shown above, the name of the clock is assumed as *A*. Clock name plays a very important role in SDC. Once a clock has been defined and given a name, all other SDC commands that depend on a clock would just refer to the name, rather than providing any other characteristics. When a clock name is mentioned, all other characteristics of the clock are known. The name provides an easier method to collectively refer to all the characteristics of the clock.

5.3.4 Specifying the Duty Cycle

The duty cycle of a clock is specified using the -*waveform* option. This option is typically an ordered pair of real numbers, representing the rising and falling edge of a clock. The numbers indicate the time when the rise and fall edge happen after time $t=0$. For example, the waveform in Fig. 5.1 has the rising edge at time $t=5$ and falling edge at time $t=10$. This can be represented as:

create_clock -period 10-name CLK -waveform {5 10} [get_ports A]

Similarly in Fig. 5.2, the rising edge is at time $t=0$ and falling edge is a time $t=7$. This can be represented as:

create_clock -period 10 -name NEW_CLK -waveform {0 7} [get_ports C]

When this option is not specified, the clock is assumed to have a 50 % duty cycle. This is equivalent to saying that the waveform is {*0 period/2*}. The numbers in the waveform option have to be monotonically increasing to represent a full period. Let us consider the Fig. 5.5; here, the clock has a low pulse with a period of *10*.

The clock edge falls at time $t=4$ and rises at time $t=5$ during its clock period cycle. Since the option can only represent rise and fall transitions in that order and the values have to be monotonically increasing, we will have to consider the

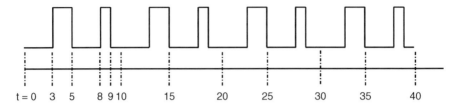

Fig. 5.6 Clock with complex waveform

clocks transition over two periods. Over a two-cycle period, the clock will have its first rising edge at $t=5$ and the next falling edge again at time $t=14$. Such a clock is represented as:

create_clock -period 10 -name CLK -waveform {5 14} [get_ports C2]

In some applications like pulse blanking, there is a need to remove data within a specified time region after a trigger. This is mainly done to reduce any RF interference. In such cases there is a need to model complex waveforms, which can be accomplished by using the *waveform* option with more than two edges. However the option must only have even number of edges representing the rise and fall transition times alternately. Let us consider Fig. 5.6, where a complex clock with period of *10*, has two pulses; the first pulse has a rise at time $t=3$ and fall at time $t=5$, the second one has rise time $t=8$ and fall at time $t=9$.

This is represented as:

create_clock -period 10 -name CLK -waveform {3 5 8 9} [get_ports C3]

As can be seen in this definition, the *waveform* option has four edges. Thus, using the *-waveform* option, we can model arbitrarily complex clock waveform.

5.3.5 More than One Clock on the Same Source

Many designs require more than one clock to be specified at clock source to meet the requirements of multiple I/O speed protocol. Let us consider the block in Fig. 5.7.

Assume that the clock port is driven from outside the block by a multiplexer on which two clocks with two different characteristics converge. In order to model the clock constraints for such a block, the designer would have to specify two distinct clocks on the same object. This is represented as:

create_clock -name C1 -period 10 [get_ports CLK]
create_clock -name C2 -period 15 [get_ports CLK] -add

Under these conditions the user would need to specify a *-add* option for subsequent clocks on the same object, if he wants both the clocks to be considered for

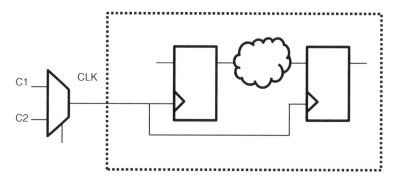

Fig. 5.7 Block driven by off-chip multiplexer with two clocks

analysis by synthesis and static timing analysis. Since each clock is required to be identified by a unique name, it is mandatory to use *-name* option, when *-add* option is used, because tool would not know how to name these two clocks (applied on the same object) distinctly. When a user specifies multiple clocks on the same object, but doesn't specify the *-add* option, the last clock definition overrides the previous definitions.

Clock is a property applied on an object with a distinct waveform and periodicity. Hence all three attributes (design object, waveform, period), together define a clock. Conceptually, a clock is different from the object (port/pin) on which it is applied. So, two different waveforms defined on the same object are two different clocks. Similarly, same waveforms and period defined on two different objects are two different clocks. That is precisely why a clock is given a unique name which refers to all the three attributes. We will observe in later chapters that this unique name of the clock is extensively used in other constraints to refer to a particular clock.

5.3.6 Commenting the Clocks

Starting SDC 1.9 a new option has been added to a few SDC commands including *create_clock*. This is called the *-comment* option. This option takes a string as its argument and is mainly used to document information about the clock to facilitate understanding, reuse, and portability of SDC and has no impact on synthesis or timing. For example,

create_clock -period 10 -name clk [get_ports clk]
-comment "Clock for USB block generated by PLL"

Some more commands have added the *-comment* option. The usage of *-comment* is the same for all the commands which support this option. This option will not be explained in subsequent sections – to prevent repetition.

5.4 Virtual Clocks

So far we have seen how to model clocks in a block. However, in some cases the user needs to constraint ports/pins in a block that has no clocks. In such cases, ports/pins are assumed to be triggered by or dependent on clocks outside the block. To capture the characteristics of such off-block or off-chip clocks, designers use the concept of virtual clocks. Virtual clocks are clocks that don't physically exist in the specific block but represent an external trigger that impacts the timing of the block. A virtual clock has no source specified. In reality, it might have a source, but that source could be outside the block being constrained. Virtual clocks are modeled using the *create_clock* command with *period*, *waveform*, and *name* option only, but source object is missing. For example,

create_clock -period 10 -name v_clk -waveform {0 5}

In Chap. 9, we will see how virtual clocks can be used in conjunction with *set_input_delay* and *set_output_delay*.

5.5 Other Clock Characteristics

Most designs require more than one clock. Having an individual clock generator for each clock is not a viable proposition. Hence this requirement of multiple clocks necessitates generation of clocks from primary clocks. When multiple clocks in a design interact, the user also has to model other characteristics like skew, latency, or phase relationship between clocks. This makes description and analysis of clocks a very complex topic. Chapters 6, 7, and 8 will cover in detail these concepts.

5.6 Importance of Clock Specification

We will see in Chap. 9 how clock specification is used to specify timing on input and output ports. However, clock specification is also very important in describing the timing requirement for the internals of the design.

For any synchronous design, there are a lot of sequential elements (flops, registers, synchronous memories, etc.), which are triggered by clocks. More than 90 % of timing paths in a design are from a sequential element to another sequential element. The path between the two flops in Fig. 5.7 shows one such path. When clocks are defined appropriately, tools can determine the time at which each of these sequential elements will trigger. This in turn will determine when the data would be launched from these elements and when these elements will capture new data. When we define a clock, this in turn defines the triggering events for thousands of sequential devices which are clocked by this clock. And, when the triggering events for the start and end point of a path get defined, the timing requirement for that path gets defined.

Usually, most paths are synchronous paths – which mean the start and end points of the path are triggered by the same root clock. Thus, as soon as we define one clock, millions of paths within a design get their timing requirement.

Let us consider two flops *F1* and *F2*, which are both triggered by the positive edge of the same clock. In Fig. 5.7, this source is the *CLK* port. Let us say, we define a clock with a period of *10ns* on this source. It immediately puts a requirement that the data launched from *F1* should reach *F2* within *10ns*.

Let us further say that *F2* is negative edge triggered. And the clock has 50 % duty cycle. Now, an active edge of *F2* will occur *5ns* after the active edge on *F1*. Thus, the timing requirement for the path is *5ns*, rather than *10ns*.

The above two examples provided a very simplified view of how clock specification defines the timing requirement for paths between two synchronous elements. In subsequent chapters, we will see how these requirements get further fine-tuned. However, this gross-level requirement still needs to be specified before further fine-tuning can take place.

5.7 Conclusion

Clocks need to be defined so that synthesis and STA consider the timing paths to the sequential elements driven by these clocks. If even a single clock is specified incorrectly, the impact could be felt by millions of paths within the design. It may cause the block to not meet timing. Even if the block meets timing, it may give a false sense of timing closure. A missing clock constraint would also mean that a huge number of paths in the design may not be timed. Since clock specifications impact maximum number of paths, even a single incorrect or missing specification could be highly detrimental to the design.

Chapter 6
Generated Clocks

Most complex designs require more than one clock for its functioning. When there are multiple clocks in a design, they would need to interact or share a relationship. Asynchronous clocks are clock signals that don't share a fixed phase relationship. Having only asynchronous clocks in the design makes it really hard to meet setup and hold requirements when multiple clock domains are interacting. We will explain about this in Chap. 7 as to why it is so. Synchronous clocks share a fixed phase relationship. More often than not synchronous clocks originate from the same source.

Today's SoCs (System on a chip) contain heterogeneous devices within the same chip. This could include very high-speed processors as well as low-speed memories all on the same chip. These elements working at different speeds are usually triggered by different clocks. Each portion operating on its own clock could bring in asynchronicity in the design. This may result in several clocks being derived from one master clock. Such clocks are referred to as *generated clocks* or *derived clocks*. These clocks can be generated in multiple ways:

1. Clock dividers
2. Clock multipliers
3. Clock gating

6.1 Clock Divider

A clock divider generates a clock of higher period and lower frequency compared to the original source clock. A typical example of a clock divider is a 2-bit ripple counter. Figure 6.1 shows the circuit of a ripple counter. For this circuit, if the period of the clock at the input of the first flop is *10ns*, then waveform generated at the LSB (least significant bit) is divided by 2, which means it has a period of *20ns*. For the same design, the waveform at the MSB (most significant bit) is divided by 4, which means it has a period of *40ns*.

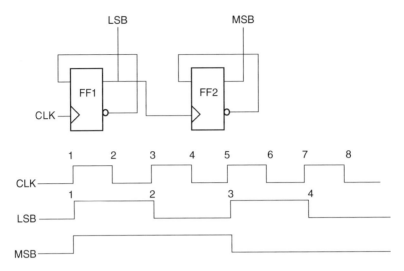

Fig. 6.1 2-bit ripple counter

6.2 Clock Multiplier

A clock multiplier is a circuit where frequency is increased and clock period is decreased for faster clocking rate. This technique is typically used in microprocessors and on internal busses so as to improve the overall throughput of the processor and is generally used in conjunction with internal cache memories. Figure 6.2 shows the circuit of a simple clock multiplier, where the clock frequency is doubled. The circuit is simple implementation of clock and its delayed version. The delay can be introduced in the line by use of buffers and invertors.

It is more common to use *PLLs* (phase-locked loops) to achieve frequency multiplication. This usage of PLLs has been mentioned in Chap. 17.

6.3 Clock Gating

Clock-gating technique has become very popular since mid-1990s to reduce power consumption. Power in a circuit is consumed when a flop or register in the design switches state due to a clock trigger. However in a design portions of the logic may not be getting used at certain times. During that stage, disabling clock to those portions of the design reduces the switching power. This is achieved by having enable logic before the clock and such a clock is called gated clock. Figure 6.3 shows the example of a gated clock.

We can also use clock gating to obtain divided clocks with waveforms similar to those shown in Fig. 6.1. The concept of clock gating can be extended to create clock pulses. Let us consider Fig. 6.4 where the clock is gated via a chain of odd number

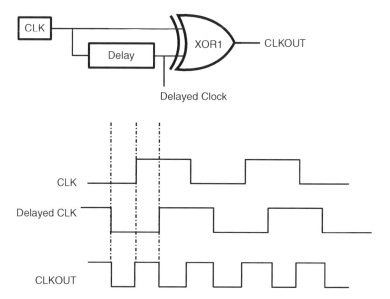

Fig. 6.2 A simple clock multiplier

Fig. 6.3 A gated clock

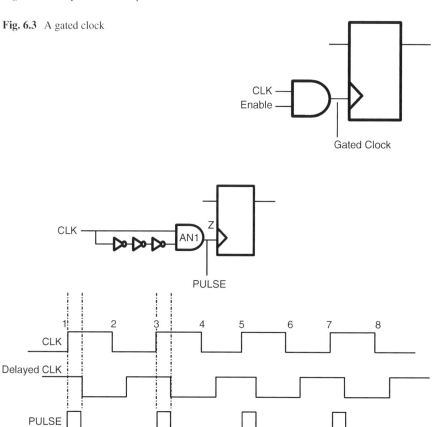

Fig. 6.4 A gated clock to generate pulse

of inverters. Depending on the delay on the chain of inverters, a pulse will be generated. This method is used to improve performance and reduce power as well. It is important to ensure that clock gating is typically done with only one clock.

6.4 create_generated_clock

The SDC command for specifying derived clocks in a design is *create_generated_clock*. The BNF grammar for the command is

create_generated_clock [source_objects]
 -*source* clock_source_pin
 [-*master_clock* master_clock_name]
 [-*name* generated_clock_name]
 [-*edges* edge_list]
 [-*divide_by* factor]
 [-*multiply_by* factor]
 [-*invert*]
 [-*edge_shift* shift_list]
 [-*duty_cycle* percent]
 [-*combinational*]
 [-*add*]
 [-*comment* comment_string]

6.4.1 Defining the Generated Clock Object

create_generated_clock is generally specified on design objects where the clock is actually available after division or multiplication or any other form of generation. These design objects called source objects can be port, pin, or net. When defining a clock on a net, ensure that net has a driver pin or the port. Otherwise the clock will not have a source. These are the points from where generated clocks can propagate into the circuit.

6.4.2 Defining the Source of Generated Clock

The source pin of a generated clock is specified using the *-source* option. This indicates the master clock source pin from which the generated clock is derived. For example, in Fig. 6.1, the generated clock is defined for LSB and MSB, and the source of the generated clock is defined at CLK.

It is better to understand the difference between a source object and the source of the generated clock. Source object refers to the point where the generated clock (or clock) is being specified, while source of the generated clock refers to the point which acts as a reference from which the generated clock has been obtained.

As indicated in Chap. 5, a source object can have more than one clock. If the master clock source pin has more than one clock in its fanin, then the generated clock must indicate the master clock which causes the generated clock to be derived. This is specified using the *-master_clock* option. This option takes the name of the SDC clock that has been defined to drive the master clock source pin. Once a generated clock has been defined, the clock characteristics (waveform, period, etc.) would be derived by the tool, based on the characteristics of the waveform at the source.

For a clock to be generated from a specific source, it is important that the source has to somehow influence the generated clock. One of the commonly committed mistakes while specifying generated clock is to specify a source which doesn't fanout to the generated clock. Effectively, this means the waveform of the generated clock has been specified as a function of the waveform at a source pin that does not even influence the generated clock! Many implementation tools do not catch this and it results in incorrect clock waveforms being used for the generated clock during STA.

6.4.3 Naming the Clock

Like the primary clock, a generated clock is also identified by its name. This is specified as string using the *-name* option. When *-name* is not specified, tools might assign a name on their own. To establish dependency on the generated clock, any subsequent SDC command simply refers to the generated clock name.

6.4.4 Specifying the Generated Clock Characteristic

The characteristic of a generated clock can be specified using one of the three options:

1. *-edges* – this is represented as a list of integers that correspond to the edge of the source clock from which the generated clock has been obtained. The edges indicate alternating rising and falling edge of the generated clock. The edges must contain an odd number of integers and should at the very minimum contain 3 integers to represent one full cycle of the generated clock. The count of edge

starts with "1" and this number ("1") represents the first rising edge of the source clock.
2. *-divide_by* – this represents a generated clock where the frequency has been divided by a factor, which means the period is multiplied by the same factor.
3. *-multiply_by* – this represents a generated clock where the frequency has been multiplied by a factor, which means the period is divided by the same factor. It should be noted that though clocks are defined using period characteristic, the multiply_by and divide_by are specified using frequency characteristic in mind (which is inverse of period).

In general any generated clock represented using *-divide_by* or *-multiply_by* options can also be represented using *-edges* option. However the vice versa is not true. Let us consider Fig. 6.1; assuming the *create_clock* is defined for *CLK*, the generated clock can be defined at *LSB* and *MSB*.

create_clock -period 10 -name CLK [get_ports CLK]
create_generated_clock -name LSB -source [get_port CLK]
-divide_by 2 [get_pins FF1/Q]
create_generated_clock -name MSB -source [get_pins FF1/Q]
-divide_by 2 [get_pins FF2/Q]

The generated clocks at *LSB* and *MSB* can also be represented using the *edges* option as:

create_generated_clock -name LSB -source [get_ports CLK]
-edges {1 3 5}[get_pins FF1/Q]
create_generated_clock -name MSB -source [get_pins FF1/Q]
-edges {1 3 5}[get_pins FF2/Q]

In the *LSB* case, the edges *{1 3 5}* indicate the edge number of the specified source clock *CLK* to which the fall and rise edges of the generated clock are aligned. For the *MSB*, since the edges are aligned to *LSB* (which is the source); hence, the edge specification is the same.

The same waveform for *MSB* can be generated using *CLK* as the source. In this case, the edge would be *{1 5 9}*.

create_generated_clock -name MSB -source [get_ports CLK]
-edges {1 5 9} [get_pins FF2/Q]

The edge specification of *MSB* depends on the edge of the source, which in this case is the primary clock *CLK*.

When a generated clock defined using *-divide_by* or *-multiply_by* options need to be inverted, then it can be specified using the *-invert* option. Let us consider Figs. 6.5 and 6.6 which have different flavors of the divide-by-two circuits. Now depending on how the generated clock is defined (inverting or non-inverting), the characteristic of the generated clock can change.

6.4 create_generated_clock

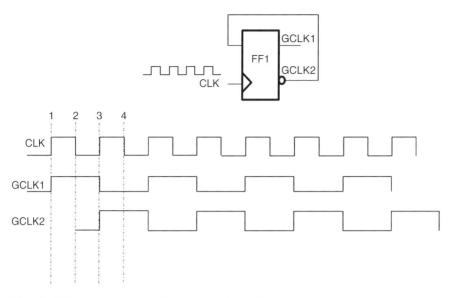

Fig. 6.5 Divide-by-two circuit with a non-inverting clock

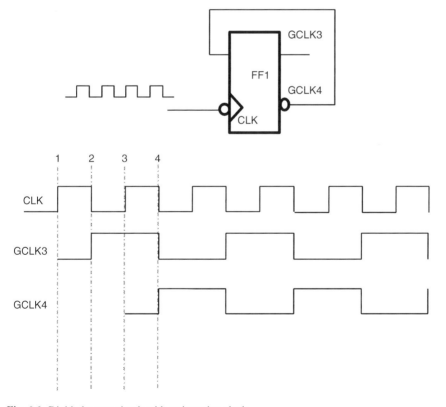

Fig. 6.6 Divide-by-two circuit with an inverting clock

In Fig. 6.5, the divider is triggered by positive edge of the source clock, and the generated clock *GCLK1* is defined as:

create_generated_clock -name GCLK1 -source [get_ports CLK]
-divide_by 2 [get_pins FF1/Q]

The clock *GCLK2* is an inverted version of *GCLK1*; this is therefore defined as:

create_generated_clock -name GCLK2 -source [get_ports CLK]
-divide_by 2 -invert [get_pins FF1/QBAR]

It should be noted that the presence of *-invert* does not change the edge of the source clock at which generated clock will have a transition. It only impacts whether the generated clock will start with a rising transition or a falling transition.

However in Fig. 6.6, the divider is triggered by negative edge of the source clock; in this case, the generated clock *GCLK3* is defined as:

create_generated_clock -name GCLK3 -source [get_ports CLK]
-edges { 2 4 6} [get_pins FF1/Q]

The clock *GCLK4* is an inverted version of *GCLK3*; this is therefore defined as:

create_generated_clock -name GCLK4 -source [get_ports CLK]
-edges { 4 6 8} [get_pins FF1/QBAR]

As it can be seen that *GCLK3* and *GCLK4* can be represented using the *-edges* option. Specifying it any other way will result in inconsistency between the actual circuit and the waveform as represented by the SDC command. This is the most commonly made mistake in defining generated clocks.

Similarly, the *CLKOUT* in Fig. 6.2 can be represented as:

create_generated_clock -name CLKOUT -source [get_ports CLK]
-multiply_by 2 [get_pins XOR1/Z]

When defining a clock where frequency is multiplied, the duty cycle can be specified using the *-duty_cycle* option. This option has meaning only with *multiply_by* option and represents the percentage of the pulse width when the multiplied clock is *1*. For example, *CLKOUT* in Fig. 6.2 can also be represented as below, indicating a 50 % duty cycle.

create_generated_clock -name CLKOUT -source [get_ports CLK]
-multiply_by 2 [get_pins XOR1/Z] -duty_cycle 50

Let us consider the Fig. 6.4, where a high pulse has been generated and the pulse width depends on the delay in the chain of the invertors. In this case, edge *1* of the clock triggers both the rising and falling edge of the pulse. This is represented as:

create_generated_clock -name PULSE -source [get_ports CLK]
-edges { 1 1 3} [get_pins AN1/Z]

Depending on the kind of pulse, the edge specification may change. For example, in Fig. 6.4, if the *AND* gate is replaced by a *NAND* gate, then it will result in a rise-edge-triggered low pulse. This would be represented as:

create_generated_clock -name PULSE_N -source [get_ports clk]
-edges { 1 3 3} [get_pins NAND1/Z]

This is because the first edge of the clock will result in a falling edge and then a rising edge of the low pulse and since *-edges* represent the order in terms of rising and falling, so it is represented as *{1 3 3}*. This implies the rising edge of the generated clock will happen due to edge *1* of the source clock. The next falling edge of the generated clock will happen due to edge *3* of the source clock followed by the next rising edge which also is on the edge *3* of the source clock. This falling edge is actually in the next pulse of the generated clock.

6.4.5 Shifting the Edges

The edges of a generated clock may need to be moved by time units to indicate shift. For example, in Fig. 6.4, if the delay through the chain of inverters is *2ns*, then the high pulse can be accurately represented as:

create_generated_clock -name PULSE -source [get_ports clk]
-edges { 1 1 3} -edge_shift {0 2 0} [get_pins AN1/Z]

The *-edge_shift* option takes a list of floating point numbers, which represents the shift in each edge in terms of time units. This option must have the same number of arguments as the number of edges to represent the shift of each edge of the generated clock. The above command now implies, on the generated clock:

Rising edge happens at the first edge of the source clock.
Falling edge happens at *2ns* after the first edge of the source clock.
Next rising edge happens at third edge of the source clock.

Similarly, for a low pulse, the representation would be

create_generated_clock -name PULSE_N -source [get_ports clk]
-edges { 1 3 3}-edge_shift {2 0 2} [get_pins NAND1/Z]

This command implies, on the generated clock

Rising edge happens at *2ns* after the first edge of the source clock.
Falling edge happens on third edge of the source clock.
Next rising edge happens at *2ns* after the third edge of the source clock.

The shift can be a positive or negative number. Use of *-edges* and *-edge_shift* can be used to model arbitrarily complex generated clocks.

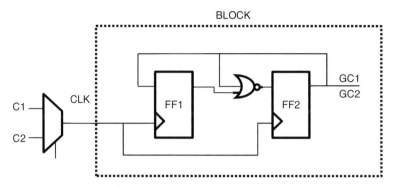

Fig. 6.7 Block driven by off-chip multiplexer with two clocks

6.4.6 More than One Clock on the Same Source

As described in Chap. 5, there can be more than one clock defined at a point. Or, for a given source, multiple clocks could be reaching the source. Typically one generated clock is defined per clock reaching the specified source. If there is more than one clock converging on the source specified for the generated clock, then the generated clock derived from this clock source pin could have characteristics corresponding to either of the clocks reaching the source. Thus, we would need to specify which of the clocks should be used to determine the characteristics of the generated clock. Let us consider the block in Fig. 6.7.

Assume that the *CLK* port is driven outside the block by a multiplexer on which two clocks with two different characteristics converge. This *CLK* port can act as a source for a clock divider circuit inside the block. In order to model the clock constraints for such a block and the generated clock for divider circuit, the designer would have to specify multiple generated clocks on the same object. This is represented as:

create_clock -name C1 -period 10 [get_ports CLK]
create_clock -name C2 -period 15 [get_ports CLK] -add

 # The following generated clock is based on C1's characteristics
create_generated_clock -name GC1 -divide_by 3 -source [get_port CLK]
-master_clock C1 [get_pins FF2/Q]

 # The following generated clock is based on C2's characteristics
create_generated_clock -name GC2 -divide_by 3 -source [get_port CLK]
-master_clock C2 [get_pins FF2/Q] -add

Thus, even though the source for both the generated clocks is the same (viz., *CLK* port), the waveform for the two generated clocks are different. Alternately,

6.4 create_generated_clock

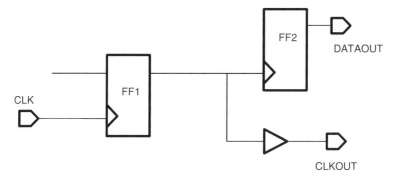

Fig. 6.8 Source-synchronous interface

there are two clock waveforms on the object *FF2/Q*. This makes sense logically. If the source of the divider has two different kinds of clocks on it, the divider output will also have two different kinds of clocks on it.

Under these conditions the user would need to specify an *-add* option, if he wants both the generated clocks to be considered for analysis by synthesis and STA. Since each clock is required to be identified by a unique name, it is mandatory to use *-name* option, when *-add* option is used. In this example for a generated clock, the specified source object had multiple clocks (either defined on it or reaching it). In such situations, it's not clear as to which of these clock characteristics should be used to create the waveform for the generated clock. The option *-master_clock* has been used to identify which of the clocks reaching the specified source object should be used for deriving the characteristics of the generated clock. When a user specifies multiple clocks (or generated clocks) on the same object but doesn't specify a *-add* option, the last constraint overrides the previous definitions.

It should be noted that *-source* uses design object on which clock is defined/ reaches, while master specifies the clock name.

6.4.7 Enabling Combinational Path

Let us consider Fig. 6.8 which represents a source-synchronous interface. In a source-synchronous interface, clock appears along with the data as an output. The advantage of this mechanism is that both clock and data are routed through similar traces and thus have very similar delays. At the receiver device, the incoming data is sampled with respect to the incoming clock. The actual trace delay is not of much importance as long as the delay differential on the two lines is close to 0. This mechanism provides an interface for high-speed data transfer.

In this figure the delay on the *DATAOUT* pin should be specified with respect to *CLKOUT*. In this case, a generated clock needs to be defined at CLKOUT. This is done using the *-combinational* option. When this option is specified, the generated clock is considered to be of the same period as master clock pin, which is equivalent to a *divide_by 1*. It cannot be used with any other option. This can be represented as:

create_generated_clock -name CLKOUT -combinational
-source [get_pins FF1/Q [get_ports CLKOUT]

In some case, there may be more than one path from the source clock pin to the place where generated clock is defined. If these paths are sequential in nature, i.e., they pass through sequential elements like flip-flops or through a transparent latch, then generated clocks are generally considered safe the way they are traditionally defined. However in some cases, if there is path from the source pin to the generated clock, which is purely combinational, that coexists with the sequential path, then traditional definitions of *create_generated_clock* will fail. In such cases, it is important to block the sequential path because the combinational path is always active. That too is achieved by defining a generated clock with *-combinational* option.

In Chap. 11 on false paths, we will see how the various kinds of clocks can be used to disable certain clock paths from timing analysis, which help in improving the efficiency of STA tools.

6.5 Generated Clock Gotchas

Since clocks can be generated in multiple ways, it is a common source of mismatch between design functionality and timing specification. While specifying generated clock, the designer must be careful about the following things:

1. If you define a generated clock make sure it is actually generated by the specified source object. Conversely, if a flip-flip or register is driven by a clock which is in fanout of another clock, make sure there is *create_generated_clock* constraint set on it. A missing generated clock may result in unconstrained registers.
2. When multiple clocks converge on the source pin of a clock, make sure to specify the master clock with the generated clock definition.
3. If you are specifying more than one generated clock constraint on a pin because of multiple sources, make sure to use the *-add* option; otherwise, the last specified constraint would override.
4. Avoid clock convergence via multiple combinational paths as it can result in a pulse. If clocks converge via multiple paths (combinational and sequential), then it is important to disable the sequential path.

6.6 Conclusion

As with primary clocks, it is important to model generated clocks correctly. Failure to do so may result in increased timing closure iterations. If the characteristic of the generated clock as defined by the SDC constraint doesn't match the actual functionality

6.6 Conclusion

of the circuit, then these are extremely difficult to debug. In many cases, the design may meet the timing, but the hardware will exhibit a totally different behavior.

When generated clocks are defined, the clock characteristics are formed based on the clock characteristic at the source. It is usually possible to define the same characteristic directly through *create_clock* on the objects, rather than using generated clocks. From timing analysis perspective, as long as the characteristics are the same, it does not matter whether the clock was specified using *create_generated_clock* or using *create_clock*. However, whenever a clock is derived from another clock, it is always better to use *create_generated_clock*, rather than *create_clock*. It is easier to maintain and enhance, as modifying the source clock characteristic will directly impact the characteristic here. Also, using the correct constraint better mimics the design intent, which reduces the chances of errors as constraints are modified or enhanced – including migration across technologies and designs.

Further, when multiple clocks in a design interact, it is not enough to simply define the clocks correctly; it is also required to correctly define relationship between clocks. In the next chapter we will cover how you can effectively define such relationship between interacting clocks.

Chapter 7
Clock Groups

When a design has more than one clock, the timing of such a design depends not just on the frequency of clocks but also on the relation the clocks share with each other. Synchronous clocks are clocks which share a deterministic phase relationship. More often than not, synchronous clocks share the same source.

On the other hand, asynchronous clocks are clocks which don't share a fixed phase relationship. Let us consider Fig. 7.1 – if the two clocks *C1* and *C2* are generated from different sources, then they are treated as asynchronous.

The section of the design driven by each of these clocks forms a clock domain. The signals that interface between these clock domains driven by asynchronous clocks are called asynchronous clock domain crossings or abbreviated as *CDC*.

In this chapter, we will understand how to specify the relation between clocks which are asynchronous in nature and how to group them into domains. But first, let us try to understand the timing impact on a design with multi-frequency clocks.

7.1 Setup and Hold Timing Check

Let us consider Fig. 7.1. In this simple circuit, there is a launch flop (*F1*) that launches data that is captured by the capture flop (*F2*). As described in Chap. 3, setup is defined as the time by which data needs to be available before the active edge of clock, and hold is the time for which the data must remain stable after the active edge of the clock, so that data is properly registered by the flip-flop.

The same concept can be extended for the design in Fig. 7.1. The design would need to ensure that data on the active edge of the launch flop (*F1*) is captured by the closest following active edge of the capture flop (*F2*). This is called the *setup timing check*. Figure 7.2 shows the waveform of the clocks for the design.

Let us assume that t_F is the delay from *Clock* to *Q* pin of launch flop (*F1*) and t_C is the delay within the combination cloud. This means data arrives at flop *F2* at time $t_F + t_C$. Let us also assume that edges of clocks *C1* and *C2* are perfectly aligned and

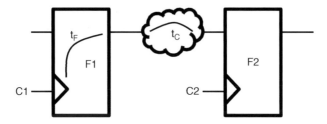

Fig. 7.1 Asynchronous clock domain crossing

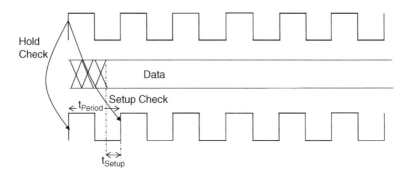

Fig. 7.2 Waveform for interacting clocks

the setup requirement of capture flop (*F2*) is t_{Setup}. Since the next clock arrives at *F2* at the next edge, which is t_{Period} (*period* of clock *C2*), then for data from flop *F1* to be captured by *F2*, the data must arrive at least t_{Setup} time before the next active edge of *F2*. This setup timing check imposes an upper bound on the timing requirement for the signal to arrive at *F2* and can be represented as:

$$t_F + t_C < t_{Period} - t_{Setup}$$

Once the setup requirement is met, for the data to be properly captured the hold requirements have to meet as well. This is measured by the *hold timing check*, which ensures the hold timing is met between the active edge of the launch clock and the same edge of the capture clock. For the same design, since $t_F + t_C$ is the time required for the data to reach flop *F2*, the time at which the data arrives must be more than the hold time (t_{Hold}) of flop *F2*, so that the current data does not corrupt the previous data. This hold timing check therefore imposes a lower bound on the timing requirement for the signal to arrive at *F2* and can be represented as:

$$t_F + t_C > t_{Hold}$$

This was a rather simple case, since we assumed clocks *C1* and *C2* had perfectly aligned edges. The equations get just a little more complicated if the edges are not aligned (though, they still originate from the same source). If t_L is the time for the

7.1 Setup and Hold Timing Check

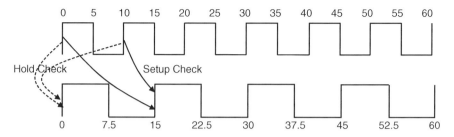

Fig. 7.3 Waveform for fast to slow clocks

clock to reach the launch flop from its source and t_z is the time for the clock to reach the capture flop from its source, then setup and hold timing check would be:

$$t_L + t_F + t_C < t_Z + t_{Period} - t_{Setup}$$
$$t_L + t_F + t_C > t_Z + t_{Hold}$$

On the other hand, if the two interacting clocks have different frequencies, then depending on their respective frequency values, the active edge of flop (*F1*) and closest following active edge of capture flop (*F2*) may vary in every clock cycle. Here are few representative examples to analyze these further.

7.1.1 Fast to Slow Clocks

For Fig. 7.1, let us consider the case when the period of the launch clock is less than the period of the capture clock. Let us further assume that *C1* has a period of *10ns* with a *50%* duty cycle and *C2* has a period of *15ns* with a *50%* duty cycle. Let the clocks be represented as:

create_clock -period 10 -name C1 -waveform {0 5} [get_pins F1/CK]
create_clock -period 15 -name C2 -waveform {0 7.5} [get_pins F2/CK]

Figure 7.3 shows the waveform of these clocks. From this it will be evident that the waveforms repeat themselves after *30ns*. Thus, any analysis has to be done only within *30ns* window. For the setup timing check, the launch/capture combinations within the window occur at

1. Launch edge at *0* and capture at *15ns*.
2. Launch at *10ns* and capture at *15ns*.
3. Launch at *20ns* and capture at *30ns*.

Out of these, the second pair is the most restrictive and is considered for setup. Similarly, if we compute all the hold check pairs within the window, we will find that the worst case combination for hold corresponds to the launch edge at *0* and capture edge at *0*. So, both edges at *0* are chosen for hold check. This ensures that data at time unit *0 at* the launch flop is not registered by capture flop at time *0*.

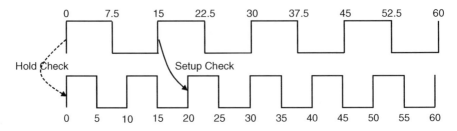

Fig. 7.4 Waveform for slow to fast clocks

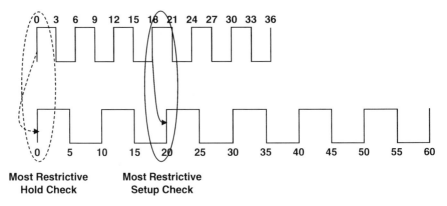

Fig. 7.5 Waveform for clocks that are not integer multiples

7.1.2 Slow to Fast Clocks

Let us look at another example of most restrictive check being used. If the period of clocks *C1* (*15*) and *C2* (*10*) are reversed, then once again all edge-pair combination till time *30ns* are considered and the most restrictive pair is used. Thus, setup timing check should be done between the launch edge at *15ns* and capture edge at *20ns*. Similarly, the most restrictive hold check is determined, which is still at time *0* for both edges. Figure 7.4 shows the waveform in this case.

7.1.3 Multiple Clocks Where Periods Synchronize in More than Two Cycles

Let us consider Fig. 7.5 where the clocks take several more cycles to realign. Let period of clock *C1* be *6ns* and period of clock *C2* be *10ns*. Assuming the clock edges are aligned at time $t=0$, the next time their edges will align will be at time $t=30$, which is the LCM of the two clock periods.

As it can be seen from the waveform, there are a number of edges where you can perform setup and hold check. But the most restrictive setup check is when launch is at *18ns* and capture is at *20ns*. Similarly the most restrictive hold check is when both edges are at *0*.

7.1.4 Asynchronous Clocks

As it is evident from these examples, these checks can get pretty complicated for multiple-frequency clocks. If the clocks don't share a phase relationship, the arrival of the launch clock and capture clock will not be deterministic relative to each other. This means setup and hold timing requirement could potentially vary in every cycle. This becomes a big timing problem when analyzing asynchronous clocks, if there is a signal in the data path driven by these clocks that may be interacting and creating an asynchronous clock domain crossing. This can potentially lead to certain issues like metastability. In Fig. 7.1, if the input of the flip-flop $F2$ is changing while it is being captured by flip-flop $F2$, then the output of $F2$ could be unstable for a certain period of time. This is called *metastability* which needs to be resolved using synchronizers. The main problem with asynchronous *CDC* is as follows: With each edge pair, there is a different timing requirement. So, at some time or other, there will be very little margin. And, since checks are supposed to be made on most restrictive pair, hence, there will be at least some edge, which will violate!

To prevent implementation tools from spending time unnecessarily to meet the timing on such paths, it is generally recommend to identify such crossings. This is achieved using *set_clock_groups*.

7.2 Logically and Physically Exclusive Clocks

Sometimes, you would have designs where clocks may not be talking to each other depending on how the design is architected. Let us consider Fig. 7.6; here the two clocks irrespective of their source don't interact with each other, even though they coexist in the design. These clocks are considered to be logically exclusive.

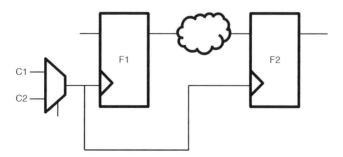

Fig. 7.6 Logically exclusive clocks (*C1* and *C2*)

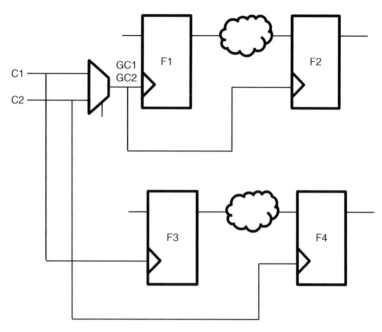

Fig. 7.7 Physically exclusive clocks (*GC1* and *GC2*)

Let us consider Fig. 7.7; here the clocks *C1* and *C2* are logically exclusive; however, the two generated clocks *GC1* and *GC2* are exclusive, but they cannot coexist together on the same net. Thus, clocks *GC1* and *GC2* are considered to be physically exclusive.

7.3 Crosstalk

When clocks are mutually exclusive, even though they don't talk, there could be interference between the signals resulting in unwanted effect. This is typically a problem seen in deep submicron technology and could be because of a number of reasons like lower geometry's requirement for higher routing density, interaction between devices, or coupling capacitance between signals. This results in a phenomenon called *crosstalk*. Let us consider Fig. 7.8.

In this figure the coupling capacitance between neighborhood nets results in unwanted and unexpected activity on the signals. This activity could be a glitch that can impact timing. The signal that is impacted is called the victim and signal that is the cause is called the aggressor. The crosstalk can affect the timing of the victim signal, if the aggressor switches at the same time as the victim. Depending upon the

7.3 Crosstalk

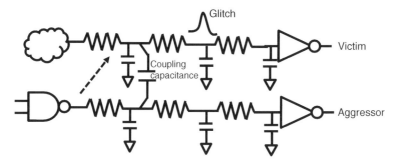

Fig. 7.8 Glitch due to crosstalk

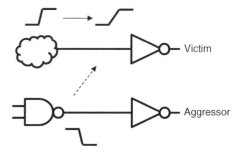

Fig. 7.9 Victim slew deterioration on account of crosstalk

direction of the switching for the aggressor and the victim, the transition at the victim could be slower (impacting setup relationship) or faster (impacting hold relationship). This is referred as the timing window relationship between the aggressor and victim and indicates the period of overlapping time when switching of aggressor and victim can potentially coincide.

Since crosstalk affects timing, it has a direct impact on setup and hold timing check. Let us consider Fig. 7.9, which is the schematic representation of Fig. 7.8 without the resistance and capacitance. If the aggressor net has a switching in the direction opposite to that of the victim, the slew on the victim net can deteriorate, thereby increasing its delay. This will impact the setup timing. Similarly, a switching on the aggressor net in the same direction as the victim can improve the slew of the adjacent victim net reducing its delay. This will impact hold timing.

From a signal integrity perspective, if mutually exclusive clocks have no crosstalk issue, then they are considered to be physically exclusive.

Most STA tools provide a way to measure the integrity of a signal in a design framework. There are books just on signal integrity and crosstalk analysis, and we will not be covering this in detail here. The concept is being introduced since certain SDC commands provide directives for crosstalk analysis.

7.4 set_clock_group

Based on what we have looked so far, correct setup and hold requirements ensure timing for reliable data capture. However for asynchronous clocks it could be tedious and impossible to meet the requirement given that the phase relationship of the clocks is not deterministic. For mutually exclusive clocks it makes no sense to try to meet the requirement, since the clocks don't talk to each other. In order to indicate to timing tools to ignore any timing paths or crosstalk analysis between asynchronous or mutually exclusive clocks, SDC provides the *set_clock_groups* command. The BNF grammar for the command is:

set_clock_groups [*-name* group_name]
　　　　　　　　[*-group* clock_list]
　　　　　　　　[*-logically_exclusive*]
　　　　　　　　[*-physically_exclusive*]
　　　　　　　　[*-asynchronous*]
　　　　　　　　[*-allow_paths*]
　　　　　　　　[*-comments* comment_string]

The *-name* option is used to provide a unique name for clock group. The clocks are divided into groups which are specified using *-group* option.

The *-logically_exclusive* option is used when clocks are mutually exclusive but can have a coupling interaction between them. The grouping between clocks in Fig. 7.6 can be represented as:

create_clock -period 10 -name C1 -waveform {0 5} [get_ports C1]
create_clock -period 20 -name C2 -waveform {0 12} [get_ports C2]
set_clock_groups -logically_exclusive -group C1 -group C2

Though the aforementioned *set_clock_groups* is technically correct, the authors recommend to create a combinational generated clock from *C1* and *C2* and then set up the clock group relation between them. This helps reuse in case the design is modified at a later stage such that clocks *C1* and *C2* start interacting in another part of the design (among *F3* and *F4*) as shown in Fig. 7.7. This would be modified as:

create_clock -period 10 -name C1 -waveform {0 5} [get_ports C1]
create_clock -period 20 -name C2 -waveform {0 12} [get_ports C2]
*create_generated_clock -name GC1 *
-source [get_ports C1] [get_pins mux1/A] -combinational
*create_generated_clock -name GC2 *
-source [get_ports C2] [get_pins mux1/B] -combinational
set_clock_groups -logically_exclusive -group GC1 -group GC2

The *-physically_exclusive* option is used when the clocks don't coexist in the design. The grouping between the clocks in Fig. 7.7 can be represented as:

create_clock -period 10 -name C1 -waveform {0 5} [get_ports C1]
create_clock -period 20 -name C2 -waveform {0 12} [get_ports C2]
*create_generated_clock -name GC1 -divide_by 1 *

7.4 set_clock_group

-source [get_pins mux1/A] [get_pins mux1/Z] -combinational
*create_generated_clock -name GC2 -divide_by 1 *
-source [get_pins mux1/B] [get_pins mux1/Z] -combinational -add
set_clock_groups -physically_exclusive -group GC1 -group GC2

As it can be seen, the timing between flops *F1* and *F2* doesn't have to be considered for the combination of *F1* being driven by *C1* and *F2* by *C2* and vice versa, but clocks *C1* and *C2* also drive flops *F3* and *F4,* and so, we cannot simply apply

set_clock_groups -logically_exclusive -group C1 -group C2

This command will disable timing paths between *F3* and *F4* for the clocks *C1* and *C2*. By defining a combinational generated clock at the output of the mux, the timing tool is given the directive to disable localized timing path analysis between flops *F1* and *F2* for the relevant clocks, without impacting flops *F3* and *F4*.

If you define multiple clocks on the same design object (using *-add* option), they should be physically exclusive. Another scenario when clocks are physically exclusive is when both system clock and test clock are applied on the same port.

The *-asynchronous* option is used when the clocks don't share a phase relationship with each other. It should be understood that asynchronous crossings also need synchronizers, purely for functional reliability. Synchronizers are not being dealt in this book, since the scope of the book is limited to timing aspects.

The options *-logically_exclusive*, *-physically_exclusive*, and *-asynchronous* are mutually exclusive. You can use only one option in a single *set_clock_groups* command. However you can specify relationships between clocks in multiple commands which could be different.

Each of these three options indicates that timing paths between clock groups must not be considered. However for crosstalk analysis, they have a different meaning. If the clock group is *logically_exclusive*, then crosstalk analysis between clocks is computed like any two synchronous clocks. If the clock group is *physically_exclusive*, then no crosstalk analysis is done between the clocks. If the clock group is *asynchronous*, the clocks are assumed to have an infinite timing window where the aggressor and victim can switch together.

When clock groups are defined asynchronous and the users want to maintain the crosstalk analysis but don't want to disable timing paths between clock, then that is achieved using *-allow_paths* option. This option can only be used with *-asynchronous* option. This is generally used only in the context of signal integrity checks and not used in STA.

You can have more than one group in a single *set_clock_groups* command. The list of clocks in a group is meant to be logically exclusive or physically exclusive or asynchronous to all the clocks in other groups. If only one group is specified, then it indicates all clocks in that group are logically exclusive or physically exclusive or asynchronous to the rest of the clocks in the design. One of the most important things to note is this command only specifies relationship between clocks in different groups. No relationship is implied for the clocks in the same group. Let us consider the command below:

set_clock_groups -asynchronous -group [get_clocks {clk1 clk2 clk3}]
-group [get_clocks {clk4 clk5 clk6}]

This command implies:

1. *clk1* is asynchronous to *clk4*, *clk5*, and *clk6*.
2. *clk2* is asynchronous to *clk4*, *clk5*, and *clk6*.
3. *clk3* is asynchronous to *clk4*, *clk5*, and *clk6*.
4. No relation can be assumed among *clk1*, *clk2*, and *clk3*.
5. No relation can be assumed among *clk4*, *clk5*, and *clk6*.

7.5 Clock Group Gotchas

While specifying the clock group the designer must be careful about the following things:

1. If you define clocks within a group, it doesn't mean they are synchronous. The relationship among clocks within a group could be defined elsewhere (say in another *set_clock_group* command or by the tool default).
2. Defining the clock group with incorrect option (*-physically_exclusive, logically_exclusive, -asynchronous*) may not impact timing since all effected timing paths are ignored, but it will impact your signal integrity analysis.
3. Just because you have defined a clock group relationship between a master clock and other clocks in the design, it doesn't mean that relationship is inherited by the generated clocks which have been derived from the master clock. All relationships should be explicitly specified.
4. The best way to remember clock grouping is
   (a) If two or more clocks coexist in the design, but there is no phase relationship, then they are specified as *-asynchronous* in *set_clock_group*.
   (b) If two or more clocks coexist in the design, but there is a circuit to select only one among these, then they are specified as *-logically_exclusive* in *set_clock_group*.
   (c) If two or more clocks cannot coexist in the design, then they are specified as *-physically_exclusive* in *set_clock_group*.

7.6 Conclusion

As much as we would like all clocks in a design to be in a single domain, the reality is multiple clock domains are inevitable. We looked at how we can ignore timing paths between domains that don't necessarily interact or which need not be timed, even if they interact. In the next chapter we will look at other clock characteristics that have to be considered for clocks.

Chapter 8
Other Clock Characteristics

In the preceding chapters, we assumed the clock to be ideal, i.e., they transition from *0* to *1* and vice versa instantaneously (have a rectangular waveform); they reach all the flops in the design at the same time (all edges align) and there is no delay between the clock generation circuit and the place where the clock is actually consumed. In reality, clocks are never ideal.

Clocks in a design form a network reaching out to a high fanout of flops. Because different flops may be at a different electrical distance from the source clock, clocks may not be reaching all the flops at the same time. Further a single clock buffer driving all the flops would see a huge capacitive load that will result in high slew. To alleviate these issues, clock tree is balanced. In this step clock buffers are inserted in the path to balance the network so as to equalize the delay to the leaf level nodes or flops. This way each buffer sees only a portion of the total load. This step is called clock tree synthesis (aka *CTS*).

Let us now try to understand how these characteristics affect the timing of a design.

8.1 Transition Time

Figure 8.1 shows the waveform of a nonideal clock. When the clock transitions from *0* to *1* or *1* to *0*, it happens over a finite period of time.

The rate of change of signal is termed as *slew*. Slew is generally measured in terms of *transition time* which is defined as the time required for a signal to change from one state to another. It is also typically measured as percentage of the total voltage change to be undergone. These markers act as the threshold settings for the measurement. For example, designer may define transition time as 30–70 % meaning, the time it takes from signal having undergone 30 % of the

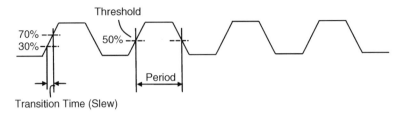

Fig. 8.1 Waveform of a non-ideal clock

voltage change to 70 % of the desired change. Rise transition time is the transition component for signal to go from *0* to *1*. Fall transition time models the opposite. Also to measure any relationship between edges of a clock, it is assumed that clock has reached its active or inactive state when it reaches a predefined threshold value (usually, 50 % of voltage level). The numbers indicated for threshold are samples and different methodologies may prescribe to a different threshold value.

Before *CTS*, because of the high load on the clock network, it doesn't make sense to do any delay calculation on the clock line as it could result in an unrealistically high *slew* value, which could affect the setup and hold time of registers. Thus, for STA this value of *slew* is assumed prior to *CTS*. This value is also specified to *CTS* tools to meet the *slew* goals. This is specified using the *set_clock_transition* constraint.

8.2 set_clock_transition

Transition time of a clock is modeled in SDC using *set_clock_transition*. The BNF (Backus-Naur Form) grammar for the command is:

set_clock_transition   [*-rise*]
                                 [*-fall*]
                                   [*-max*]
                                   [*-min*]
                                   clock_list
                                   transition_time

The *-rise* option is used to provide the transition time for the rising edge of clock. Similarly *-fall* is used to provide the transition time for the falling edge of clock. The options *-max* and *-min* model transition time for maximum and minimum operating conditions. The options *-rise*/*-fall*/*-max*/*-min* can be used separately or in tandem. However care must be taken not to contradict the transition time. For example, transition time cannot be negative or the max value should not be less than the min value. The transition time can be set on one clock or set of clocks. The specified transition time is applicable for the clock network, and the same value is used for the clock pin of the sequential element in the network of the clock. The clock used in this constraint should match the name of the clock used in *create_clock* or *create_generated_clock* constraint.

 # Rise transition on Clock C1
set_clock_transition -rise 0.2 [get_clocks C1]

 # Fall transition on Clock C2 for min and max conditions
set_clock_transition -fall -min 0.2 [get_clocks C2]
set_clock_transition -fall -max 0.4 [get_clocks C2]

 # Transition (rise, fall, min, max) on all clocks in the design
set_clock_transition 0.3 [all_clocks]

It should be noted that *set_clock_transition* is to be used only during pre-layout stage, before clock tree synthesis has been done. This command should never be used for any post-layout timing analysis, after the clock tree has been synthesized.

It should be understood that the transition value specified by this command is the time taken to transition from one state to another. However, the threshold itself for measurement of the transition time is a property of the characterization library.

8.3 Skew and Jitter

When a clock is generated by a source, it may not arrive at all the flops at the same time. The difference in the arrival time at various flops could be because of different paths through clock network, or coupling capacitance or other *PVT* (Process, Voltage, Temperature) variations in the design. This causes the edges of the same clock not to align when they reach the various devices. This difference between clock arrivals at different points in the design is referred to as clock *skew*. Clock skew can be between different points of the same clock (*intraclock*) or different (usually, synchronous) clocks (*interclock*).

At the clock generating device (say: PLL) itself, a clock's edge may not be deterministic on account of crosstalk or electromagnetic interference or due to PLL characteristics. This undesired deviation in the periodicity of a clock is referred to as *jitter*. Because of skew or jitter issues, design can have setup and hold violation. As shown in Fig. 8.2, these deviations cause the clock edge to slide on the time scale thereby reducing or increasing the time available for meeting the setup and/or hold requirements. Skew and Jitter cause lack of predictability as to when will an exact edge arrive at the point of the trigger (sequential device). These are called *Uncertainty*.

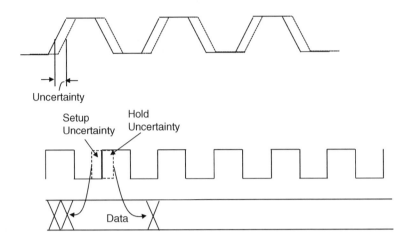

Fig. 8.2 Impact of uncertainty on setup and hold

8.4 set_clock_uncertainty

Clock skew and jitter are modeled in SDC using *set_clock_uncertainty*. The BNF grammar for the command is:

set_clock_uncertainty [*-from* | *-rise_from* | *-fall_from* from_clock]
  [*-to* | *-rise_to* | *-fall_to* to_clock]
  [*-setup*]
  [*-hold*]
  [*-rise*]
  [*-fall*]
  [object_list]
  Uncertainty_value

8.4.1 Intraclock Uncertainty

When modeling *skew* or *jitter* on a single clock (intraclock uncertainty), you need to specify the name of the clock, port, or pin. When you specify a clock, it means the uncertainty applies to all sequential elements driven by the clock. When applied on a port or pin, it applies to all clocks (and their corresponding sequential elements) in the fanout of the port or the pin.

The user can also specify different uncertainty values for setup and hold checks using the *-setup* and *-hold* options. For intraclock uncertainty, setup is impacted by both jitter and skew. While hold is impacted only by skew, not jitter.

A hold check is made at the same edge for launch and capture clock. Thus, any jitter impacts both the launch and the capture devices by exactly the same amount, and in the same direction. Thus, intraclock hold analysis need not care about jitter. Hence, it is recommended to have different values for setup and hold uncertainty with hold (only skew) being less than setup (skew and jitter).

 # Intraclock Uncertainty
set_clock_uncertainty 0.5 [get_clocks C1]
set_clock_uncertainty -setup 0.5 [get_clocks C2]
set_clock_uncertainty -hold 0.2 [get_clocks C2]

In the example above, if there are paths from other clocks to *C1*, then unless otherwise specified, the intraclock uncertainty also models the interclock uncertainty. However the user can explicitly state the interclock skew as described in Sect. 8.4.2.

8.4.2 Interclock Uncertainty

While modeling interclock uncertainty, source (start) clock is specified using the *-from* option and the destination (end) clock is specified using *-to* option.

If a designer wants to model uncertainty to be different for rise and fall edges, then he should use the options *-rise_from, -fall_from, -rise_to, -fall_to*. These options were added to SDC as an afterthought. Prior to these options the user could specify them using *-rise* and *-fall* option. However that didn't provide a very fine granularity, e.g., you could want to set uncertainty for rise condition on *-from* and fall condition on *-to*. It may be matter of time before these may be removed from the standard. Most STA tools treat this as an obsolete option today.

If the user wants to model the uncertainly only for setup checks or wants to use different values for setup and hold, they can use the *-setup* and *-hold* option. Setup would now need to include the jitter and skew for both the clocks. And unlike intraclock, in this case jitter and skew for both clocks need to be taken into account for hold as well, because both launch and capture clocks could have their own skew and jitter; and in worst case scenario, these could be in opposite directions, e.g., for setup, the launch clock could be delayed, while the capture clock could be earlier; and for hold, the launch clock could be early, while the capture clock could be delayed.

The following commands will apply the uncertainty value to all flops and latches that are driven by the respective clocks.

 # Clock uncertainty from C1 to C2 for setup and hold
set_clock_uncertainty -from C1 -to C2 -setup 0.5
set_clock_uncertainty -from C1 -to C2 -hold 0.5

 # Clock uncertainty from rising edge (C1) to falling edge (C2)
set_clock_uncertainty -rise_from C1 -fall_to C2 0.5

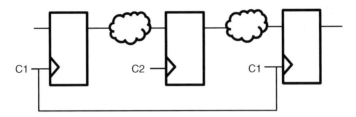

Fig. 8.3 C1 to C2 and C2 to C1 paths

Let us consider the circuit as shown in Fig. 8.3. Here, there is path from *C1* to *C2* and also a path from *C2* to *C1*. One of the more commonly made mistakes is to set the uncertainty from *C1* to *C2*. This will not cover the path from *C2* to *C1*. Uncertainty must be specified to cover all source and destination clock combinations.

 # Clock uncertainty between C1 and C2
set_clock_uncertainty -from C1 -to C2 0.5
set_clock_uncertainty -from C2 -to C1 0.5

When both interclock and intraclock uncertainties are specified, then interclock takes a higher precedence.

 # Intraclock and Interclock Uncertainty Conflict
set_clock_uncertainty 0.6 -from C2 -to C1
set_clock_uncertainty 0.5 [get_clocks C1]

In the example above the two uncertainty values on Clock *C1* are different. For all paths from *C2* to *C1*, the interclock uncertainty takes precedence, and a value of *0.6* is used. For all other paths where the destination clock is *C1*, the intraclock value of *0.5* is used.

Another important aspect of uncertainty is that its value varies between pre-layout and post-layout. In the pre-layout stage there is no CTS performed; the uncertainty value must take into effect the possible impact of skew that will be inserted. However post CTS, the skew portion is already known and doesn't need to be specified as uncertainty. So, the clock uncertainty in the post-layout stage is generally less than pre-layout.

The following table summarizes which of the factors (among skew and jitter) should be contributing to which kind of uncertainty:

	Intraclock (same clock on source and destination)		Interclock (source and destination having different clocks)	
	Setup	Hold	Setup	Hold
Pre-layout	Skew and jitter	Skew	Skew and jitter	Skew and jitter
Post-layout	Jitter	X	Jitter	Jitter

8.5 Clock Latency

Let us consider Fig. 8.4a, b, which shows delay in the clock path. The delay between the source of the clock and the actual pin where clock is used to trigger a device is defined as *clock latency*. This delay is because of the capacitive load on the interconnect or elements in the clock tree between the clock source and the clock pin. Clock latency has two components – source and network latency. *Source latency* is the delay from the source of the clock to the point where clock is defined (in SDC, through *create_clock/create_generated_clock*). This source could be on-chip or off-chip. *Network latency* is the time it takes for clock to propagate

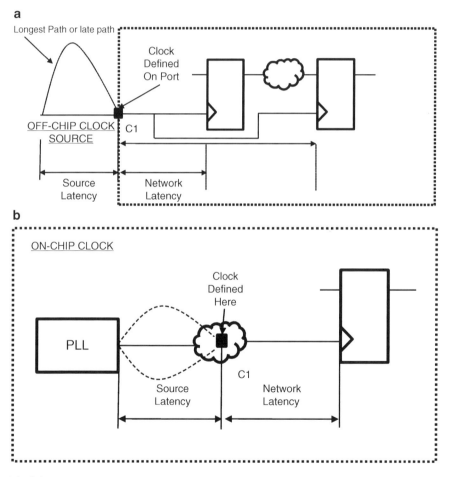

Fig. 8.4 (a) Delay on clock path with off-chip clock source (b) Delay on clock path on-chip clock source

from the point where clock is defined to the point where it is actually used to trigger the sequential device. There could be more than one clock path to a device, in which case the delays along the path could be different as well. The longest path or the one which has maximum delay is often referred to as the *late path* and shortest path or the one which has minimum delay is referred to as the *early path*. The total latency is the sum of source and network latency. Depending on the phase of the design flow you are, you can compute it or make an approximation. However accurate delay in a circuit is possible only when you know the complete implementation details – including, placement, wire connections, parasitic etc. (on account of the placement of the buffers) on the line, which is typically available post-layout. Hence to model these network delays before Clock Tree Synthesis or pre-layout stage, designers use the SDC constraint *set_clock_latency*. *set_clock_latency* is used to model delay in the clock network.

8.6 set_clock_latency

The BNF grammar for the command is:

set_clock_latency   [*-rise*]
          [*-fall*]
          [*-min*]
          [*-max*]
          [*-source*]
          [*-late*]
          [*-early*]
          [*-clock* clock_list]
          delay
          object_list

The options *-rise/-fall* specify the latency on the rising and falling edge of the clock. The options *-min/max* specify the latency for minimum and maximum operating conditions. The options *-rise/-fall/-max/-min* can be used separately or in tandem. However care must be taken not to contradict the latency values. For example, latency cannot be a negative value or the max value should not be less than the min value.

To specify the source component of the latency the user needs to provide the *-source* option along with the name of the clock. Let us consider Fig. 8.4a, assuming the SDC clock defined at port is *C1*, then source latency is used to model the delay from the off-chip source to the port. This is represented as:

set_clock_latency -source 0.5 [get_clocks C1]

Let us consider Fig. 8.4b. If there are multiple paths from the *PLL* to the place where clock is defined (output of the PLL), then the delay on the longest

8.6 set_clock_latency

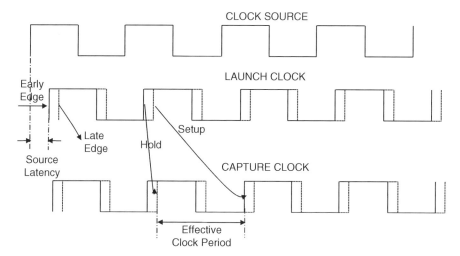

Fig. 8.5 Setup and hold check with source latency

path is modeled using *-late* option. The delay on the shortest path is modeled using *-early* option.

set_clock_latency -source -early 0.5 [get_clocks C1]
set_clock_latency -source -late 1.0 [get_clocks C1]

Clock source latency impacts the way setup and hold check may be performed when multiple clocks are interacting. To be pessimistic it is generally recommended to do setup check between the late version of the source clock and early version of the destination clock and hold check between early version of the source clock and late version of the destination clock. Figure 8.5 shows the setup and hold relationship between multiple clocks when latencies are involved.

When *-source* option is not specified, the command models the network latency. As explained earlier, *CTS* inserts buffers to distribute the huge fanout on the clock network. This is also done in a way to ensure delay on all the paths in the clock network are kept similar. These additional buffers will result in some delay on the clock path. This delay is modeled as network latency. So, effectively, network latency means delay through the clock network, which comes into being mostly because of clock tree.

This can be applied on the clocks or ports and pins. For Fig. 8.4a, the following command indicates the delay in the network of the clock. This delay applies to all the sequential devices triggered by this clock.

 # Network Latency – Applies to rise (for max and min conditions)
set_clock_latency 0.5 -rise [get_clocks C1]

 # Network Latency – Applies to fall (for max and min conditions)
set_clock_latency 0.3 -fall [get_clocks C1]

When applied on ports or pins, it implies the latency is till the clock pins of the registers which are in the fanout of these design objects. If you want to apply a specific latency value to only a part of a clock tree, it can be achieved by specifying the latency on a pin, such that the portion of the tree falls in the fanout cone of the pin. Along with port and pins, the user can also specify the associated clock using the *-clock* option. This is used when multiple clocks pass through a port or pin.

```
# Network Latency to all register clock pins in fanout of A
set_clock_latency 0.5 [get_ports A]
```

```
# Network Latency to all register pins in fanout of B, which are clocked
# by C1 and C2
set_clock_latency 1.0 -clock {C1 C2} [get_ports B]
```

There is a conceptual difference between source and network latency in SDC. Source latency can only be applied on clocks, while network latency can be applied on clock, ports, and pins. Source latency can only be used to model early and late paths. Finally, network latency is an estimate of delay prior to clock tree synthesis and is not specified after *CTS*. After CTS, it is recommended to use *set_propagated_clock* command to give directive to the tool that clock network latency needs to be computed based on the actual circuit elements – including parasitics. The actual network latency after CTS is also referred to *insertion delay*. That said, source latency still need to be specified after *CTS* for the propagated clock.

8.7 Clock Path Unateness

As clock propagates through the design, it has to pass through combinational elements. When it passes through buffer or gates like *AND/OR*, its sense (viz: direction of transition) is preserved. When it passes through inverters or gates like *NAND/NOR*, its sense is inverted. In either case, it is possible to figure out the sense of the clock, along its path. Such a clock where based on its propagation, you can figure out the sense of the arriving clock edge at flip flops is said to be *unate*. A *positive unate* clock is one where a rising edge at the source of a clock results in a rising edge at the clock pin of the flop. A *negative unate* is one where a rising or falling edge of at the source of a clock results in falling or rising (just the opposite) at the clock pin. Figure 8.6 shows example of a positive and negative unate clock.

However in some cases, depending on the circuit it may not be possible to figure out the sense. Such a clock path is said to be *non unate*. Figure 8.7 shows the example of a non-unate clock.

For such clock paths, a user can specify the *set_clock_sense* command to pick which sense (positive or negative) should propagate on the path.

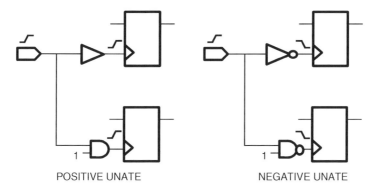

Fig. 8.6 Positive and negative unate clock

Fig. 8.7 Non-unate clock

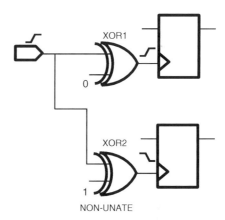

8.8 set_clock_sense

The BNF grammar for the command is:

set_clock_sense [*-positive* |*-negative* |*-stop_propagation*]
              [*-pulse* pulse]
              [*-clock* clock_list]
              pin_list

This command only is meaningful in the context of a non-unate clock network. To specify the unateness of the clock network, the user needs to specify the name of the design pins through which the clock passes. For example, to propagate only the negative clock for Fig. 8.7, the SDC can be written as

set_clock_sense -negative [get_pins XOR2/Z]

If multiple clock paths converge on such an *XOR* gate, then aforementioned command would imply that all negative edges of the clocks would be considered for analysis. To selectively apply this on certain clocks use the *-clock* option. For example,

set_clock_sense -positive -clock [get_clocks C1] [get_pins XOR1/Z]
set_clock_sense -negative -clock [get_clocks C2] [get_pins XOR2/Z]

Mutually exclusive to the *-positive/-negative* option is the *-stop_propagation* option. This is used, if designer wants to disable the propagation of certain clocks.

*set_clock_sense -stop_propagation -clock [get_clocks C3] *
[get_pins XOR1/Z]

So when do you use *-stop_propagation*? Let us consider an example where a *PLL* has multiple outputs, each output represents a clock. Sometimes clocks may be for different modes. These clocks from *PLL* output fanout to all the elements in the design, but in a certain mode you don't want a clock to reach to a particular set of flops. In that case, you would want to stop the clock propagation before it reaches the flops. This is achieved using the *stop_propagation* option. We will read more (in Chap. 14) about mode-based analysis, and that should explain why in some situations, some clocks might have to be stopped from reaching at a few points, in certain specific modes.

In Chap. 6 (Sect. 6.4.4, Fig. 6.4), we had shown how to model a pulse using *create_generated_clock*. Another way to model it is using the *-pulse* option in *set_clock_sense* as shown below.

set_clock_sense -pulse rise_triggered_high_pulse [get_pins AN1/Z]

The *-pulse* option can take one of the four values: *rise_triggered_high_pulse, rise_triggered_low_pulse, fall_triggered_high_pulse, fall_triggered_low_pulse*. A rise triggered pulse is one where the clock source has a rising active edge, whereas in a fall triggered it has falling active edge. Figure 8.8 shows the different kinds of pulse waveforms.

The advantage of using *set_clock_sense* for a pulse over *create_generated_clock* is that it doesn't create an additional clock and hence an additional domain. For a pulse, the width of the pulse is assumed to be zero, no matter what method you use. However when only *set_clock_sense* is used, the actual width of the pulse is calculated based on the differential values of rise and fall latency. The user can set the latency on the pin where pulse is generated. For example, the aforementioned *set_clock_sense* will be complemented by

set_clock_latency -rise 0.2 [get_pins AN1/Z]
set_clock_latency -fall 0.9 [get_pins AN1/Z]

This will result in a pulse of width *0.7ns*. In general the pulse width is calculated as

Pulse_width = | rise_latency -fall_latency |

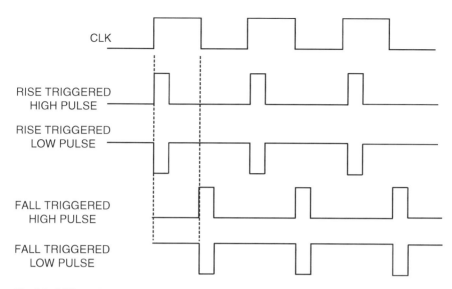

Fig. 8.8 Different kinds of pulse waveforms

8.9 Ideal Network

In preceding sections we saw how to create non-ideal clocks to take into account the impact of slew, skew, jitter, and latency. However in a design there may be a need to define certain points as ideal. This is required specifically when user wants to adopt a specific methodology to synthesize a particular portion of the design. For example, high fanout nets like scan, if taken through the traditional synthesis flow, may cause the tools to spend unwanted time trying to optimize the design to meet timing. In some cases, it may optimize design objects away, which may not be the intent. To prevent this, user can define design objects like cells, pins, or nets as ideal, which implies that such objects are not required to adhere to design rules like maximum capacitance, fanout, and transition. Also, on sources that drive such objects optimization will not be done and timing will not be updated on paths to these objects. This network of ideal cells, pins, and nets is called an ideal network and this is modeled using *set_ideal_network* constraint.

The BNF grammar for the command is:

set_ideal_network   [-*no_propagate*]
                      object_list

This can be set on ports, cells, or nets at any level of hierarchy. It can be set on any internal pin of the design, but cannot be set on a pin at a hierarchical boundary. When this constraint is set, then all objects (net, pins, cells) in the transitive fanout of the source are also considered ideal. Typically ideal net propagates through all combinational elements and stops at a sequential element. To prevent any propagation through combinational elements, the -*no_propagate* option is used.

By default, the transition and latency of an ideal network is assumed to be zero. However these can be set to specific values using the *set_ideal_transition* and *set_ideal_latency* constraints.

8.10 Conclusion

In last four chapters we studied different clock characteristics and how to model them. Constraining the clocks is the most imperative step in timing. It forms the basis of timing analysis.

The constraints learnt in this section are especially interesting. These constraints change their values and shapes between pre-layout and post-layout. Some constraint which was used during pre-layout should be replaced by another constraint during post-layout stage; or, sometimes the value might need to be modified.

We will now understand how to constrain the remaining ports of the design.

Chapter 9
Port Delays

Once the clock constraints have been applied, all the register to register paths can be timed. Now the delay constraints have to be applied on non-clock ports. If input and output port constraints are not specified, timing analysis tools assume a highly optimistic timing requirements on the interfaces. They assume the combinational logic inside the block can have the entire period to itself and leave nothing for the portion of the signal outside the block.

9.1 Input Availability

For each input port, we need to specify the time at which the inputs would be available. Consider the circuit shown in Fig. 9.1. For the block *B1*, we need to know the time at which the signal arrives on input *I1*.

This tells the implementation tools the amount of time that can be spent in the combo cloud, *C1* – between the input port and the first register element. This in turn decides the level of optimization required. This also allows the STA tools to report if the signal would reach *F1* – the first register after crossing the combinational logic – in time to be latched reliably. Consider the circuit shown in Fig. 9.2.

For the block *B2*, let us consider the input *I1* for which we need to specify the input arrival time. The signal will start at *F1* (of the previous block *B1*) when the flop gets a clock trigger. So this clock trigger acts as a reference event. After this reference event, the signal has to travel through the flop (*Clk_to_Q* delay), the combo logic *C1* (in the previous block *B1*), the combo logic *C2* (in the top level), and the interconnect delays through the wires. However, as far as the signal arrival time at *I1* is considered, it does not matter how the delay is distributed between *C1*, *C2*, *Clk_to_Q*, interconnect, etc. It just needs to know how much time after the reference event the signal appears at the input port. A late arriving signal means a very small amount of delay in the logic cloud *C3* (between the input port and the register).

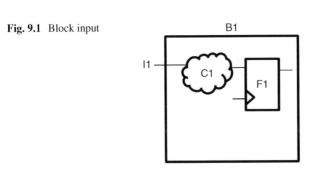

Fig. 9.1 Block input

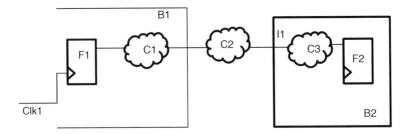

Fig. 9.2 Input available time

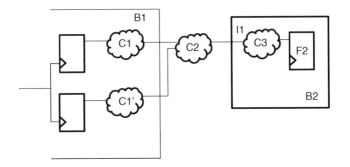

Fig. 9.3 Multiple paths from same reference event

9.1.1 Min and Max Availability Time

It is not always possible to specify the exact time at which the signal would be available at an input port. There could be multiple paths from the same reference event as shown in Fig. 9.3, or even PVT variations could cause some degree of uncertainty as to when the signal will reach the input port.

9.1 Input Availability

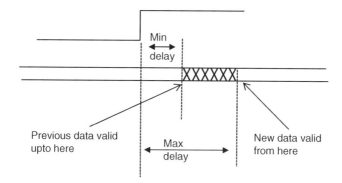

Fig. 9.4 Data valid window

The designer needs to specify the earliest time at which a signal can change at the input port. This also means the minimum time, before which the input signal would not be reaching the input port. By corollary, this means that the previous value would be held at the input port till this time. This minimum value is useful for ensuring that the input signal will not violate the hold requirement on *F2*.

The designer needs to also specify the maximum time, within which the input signal would surely be available at the input port. This also means the latest time, within which all changes to the signal would be available at the input port. This maximum time is used to ensure that this signal meets the setup requirement of the flops inside *B2*. Figure 9.4 shows the impact of minimum and maximum delay on the data validity window.

9.1.2 Multiple Clocks

Sometimes, an input signal might be triggered by multiple clocks. Consider the circuit shown in Fig. 9.5.

Signals reaching *I1* could be generated either from *Clk1* (in block *B1*) or Clk2 (in block *B2*). Both these triggering events could be independent. In such a case, the arrival time has to be specified with respect to both the reference events. It is the responsibility of the implementation tools and the STA tools to consider each of these arrival times independently and try to satisfy both the conditions.

9.1.3 Understanding Input Arrival Time

Looking back at the circuit in Fig. 9.1, let us consider that the earliest time at which a signal is available at *I1* is *3ns*. Let us say that the minimum delay for the combo cloud C1 is *4ns*. So the earliest that the signal can be available at the flop *F1* is at *7ns*.

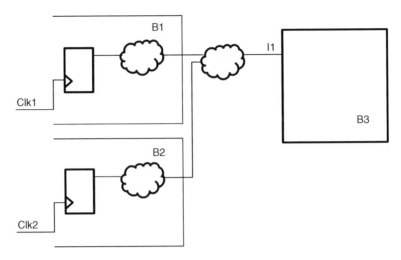

Fig. 9.5 Multiple reference events

Thus, as long as the hold requirement of the flop is less than *7ns*, the new value does not interfere with the capture of the previous data.

Similarly, let us consider that the latest time for the signal to arrive at *I1* is *5ns*. Let us also say that the maximum delay for the combo cloud *C1* is *6ns*. So the latest that the signal can be available at the flop *F1* is *11ns*. Let us further assume that the setup requirement of the flop *F1* is *0.5ns*. Thus, as long as the clock reaches *F1* at time *11.5ns* or later, the current value can be captured by the flop reliably.

9.2 Output Requirement

For each output port, we need to specify the time for which a signal travels outside the block, before getting sampled. Consider the circuit shown in Fig. 9.6. For the block *B1*, we need to know the time that is needed by the signal to travel after emerging out of *O1*, before getting sampled.

This tells the implementation tools how much logic can be put in the combo cloud, *C1* – between the last register element (*F1*) and the output port *O1*. This also allows the STA tools to report if the signal would be available at *O1* at a time – such that it still is left with sufficient time required to travel outside the block, after emerging out of *O1*. Consider the circuit shown in Fig. 9.7.

For the block *B1*, let us consider the output *O1* for which we need to specify the output required time. After coming out of *O1*, the signal gets sampled by the flop *F2* in block *B2*. The clock that triggers *F2* acts as a reference event. Before this reference event arrives on *F2*, the signal has to travel through the combo logic *C2* (in the top level), the combo logic *C3* (in the next block), and the interconnect delays through the wires. And the signal needs to still reach *F2* slightly before the reference event, so that the flop meets its setup requirement. However, as far as the signal

9.2 Output Requirement

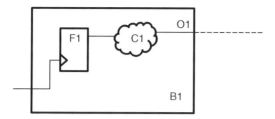

Fig. 9.6 Block output

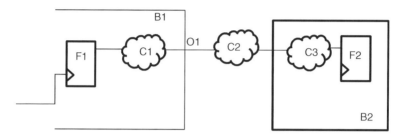

Fig. 9.7 Output required time

required time at *O1* is considered, it does not matter how the delay is distributed between *C2*, *C3* *F2*'s setup, interconnect, etc. It just needs to know how much time before the reference event the signal needs to be available at the output port. A higher time outside *O1* means a very small amount of delay in the logic cloud *C1* (between the register and the output port).

It is to be understood that the output requirement is specified in terms of how much time more is needed outside the block, before the signal gets sampled. It is not specified in terms of when the signal needs to be available at the output. That can be calculated. For example, if an output delay is specified as 6, that means the signal needs *6ns* after coming out. And if the signal is to be sampled by a clock trigger at *10ns*, that would mean the signal should be available at *4ns*. In the world of SDC, the user specifies *6* – the time that is needed after coming out. Whoever needs to know the time when the signal should be available can compute that value. This is a conceptual difference between specifying input delay and output delay. At the input, the value specified directly gives the time when the input would be available. At output, the value specified says for how long the signal will travel further. This is different from when the signal needs to be available at the output.

9.2.1 Min and Max Required Time

It is not always possible to specify the exact time for which the signal would need to travel after coming out of an output port. There could be multiple paths from the

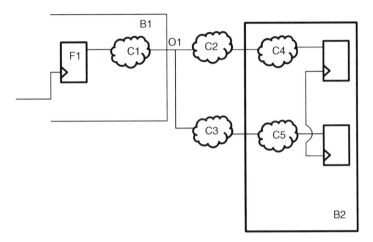

Fig. 9.8 Multiple paths to same reference event

output port to the same reference event as shown in Fig. 9.8, or even PVT variations could cause some degree of uncertainty as to when will the signal reach the output port.

The designer needs to specify the minimum time that the signal needs to travel after the output port. The designer needs to also specify the maximum time that the output signal might need to travel after coming out of the output port.

9.2.2 Multiple Reference Events

Sometimes, an output signal might be captured by multiple reference events. Consider the circuit shown in Fig. 9.9.

Signals from *O1* could be captured either by *Clk1* or *Clk2*. Both these capturing events could be independent. In such a case, the required time needs to be specified with respect to both the reference events. It is the responsibility of the implementation tools and the STA tools to consider each of these required times independently and try to satisfy each of the requirements.

9.2.3 Understanding Output Required Time

Looking back at the circuit in Fig. 9.6, let us consider that the minimum time that the signal takes outside *O1* is *3ns*. So, even if the delay inside the block till the output port O1 is *"-3ns,"* the final delay at the end of the path is zero. So the new value would still not interfere with the previous data, because that is available till the clock edge. Thus, the previous value should be held stable at *O1* till time *"-3ns."* This can be thought of as a situation, wherein *O1* is feeding into a hypothetical flop,

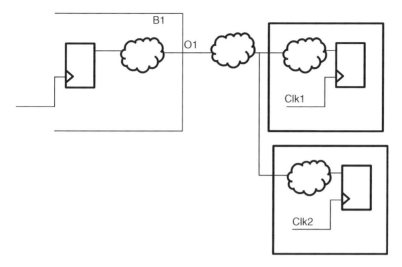

Fig. 9.9 Multiple reference events

which has a hold requirement of "*-3ns*." Note the negative sign! So a minimum delay specified for the output port is equivalent to a hold check on a hypothetical flop, where the hold value to be checked is negative of the delay value specified.

Similarly, let us consider that the maximum required time for the signal *O1* is *7ns*. That means that the signal needs to travel for a further time of *7ns*, before getting captured by the next flop. This can be thought of a situation, wherein *O1* is feeding into a hypothetical flop, which has a setup requirement of *7ns*. Assuming a clock period of *10ns*, the maximum combinational delay for the cloud *C1* can be *3ns (10–7)*.

9.3 set_input_delay

The SDC command for specifying delays on input ports is *set_input_delay*. The BNF grammar for the command is:

set_input_delay [*-clock* clock_name]
[*-clock_fall*]
[*-level_sensitive*]
[*-rise*]
[*-fall*]
[*-max*]
[*-min*]
[*-add_delay*]
[*-network_latency_included*]
[*-source_latency_included*]
delay_value port_pin_list

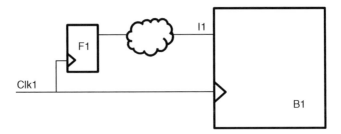

Fig. 9.10 Clock specification for input delay

9.3.1 Clock Specification

-clock option is used to specify the reference clock, with respect to which the delay value is specified. This should usually refer to the name of the clock which is used to trigger the signal that reaches this input port. Consider the circuit shown in Fig. 9.10.

Let us assume that the input *I1* of the block *B1* is being driven by flop *F1*. The flop *F1* lies outside *B1*. It could be lying in some other block, or it could be part of the top level logic. Let us further assume that the clock which triggers *F1* also goes into the block *B1*. And the name given to this clock in *B1* is *CLK1*. In that case, the *clock_name* specification should be *CLK1*. The input arrival time is triggered by this clock – which is the reference. It does not matter what is the name of the same clock signal when it drives *F1*. For example, the same clock signal could be named *CLK_1* for the portion which contains *F1*. But, for input delay of *B1*, the name has to be specified in terms of the clock that is seen by *B1*.

Sometimes, the clock signal which triggers *F1* might not enter the block *B1*. In such a situation, a virtual clock needs to be declared for *B1*. The characteristic of this virtual clock has to be exactly the same as the clock which triggers *F1*. Now, this virtual clock can be specified as the *clock_name* with the *-clock* option.

By default, it is assumed that the delay specified is with respect to the positive edge of the clock mentioned. However, if the flop *F1* is triggered by negative edge of the clock, then option *-clock_fall* needs to be added. Presence of this option will cause the reference event to be the negative edge of the specified *clock_name*, instead of the positive edge.

Consider a clock with period *10ns* and having a waveform of *{0 5}*. Let us say that the flop *F1* is triggered by falling edge of the clock. So we could specify a delay of 2 with respect to the falling edge of the clock, or we could specify a delay of 7 with respect to the rising edge of the clock. In terms of timing analysis, both the following commands will have the same impact:

set_input_delay -clock CLK1 -clock_fall 2.0 [get_ports I1]
set_input_delay -clock CLK1 7.0 [get_ports I1]

9.3 set_input_delay

However, if the launching flop *F1* is actually triggered by the falling edge of the clock, then we should use the first command, because it is a better reflection of the design circuitry.

If the input pin being considered is part of a combinational only path, then there is no clock which triggers the signal arriving on it. In such a case, *-clock* need not be specified. For such *set_input_delay* commands, the reference event is considered as time 0. If *-clock* is not specified, *-clock_fall* has no significance. Since most designs today are synchronous, hence, usually, *set_input_delay* has clocks specified. Purely combinational paths – even if existing – may also be constrained using other constraints (discussed in Chap. 13).

9.3.2 -level_sensitive

If the launching device is a latch, rather than a flop, this switch should be specified. Conceptually, use of this switch allows for the condition that the launching latch could be borrowing the time from this cycle. That means the setup slack could reduce – to account for signal starting from the latch anytime when it is transparent.

However, this option should be used after a very careful consideration. Different tools are known to treat this option differently. Some tools and translators simply ignore this option. If a designer intends to use this option, he/she should make sure that all tools in his/her flow treat this option uniformly.

9.3.3 Rise/Fall Qualifiers

-rise is used to qualify that the input delay corresponds to the signal rising at the input port, and *-fall* is used to qualify that the input delay corresponds to the signal falling at the input port. The command needs to provide the rise or fall qualifiers, if the input arrival times are different for a signal which is rising at the input port and a signal which is falling at the same input port.

When the *-rise* or *-fall* qualifier is not specified, the given value applies to both kinds of transitions. In CMOS circuits, typically path delays for rise and fall transitions are quite similar. So *-rise/-fall* specifications are not used that often.

9.3.4 Min/Max Qualifiers

-min is used to qualify that the delay value specified corresponds to the earliest arrival time for the signal at the input port. This value is used for performing hold checks inside the current design. *-max* is used to qualify that the delay value

specified corresponds to the latest arrival time for the signal at the input port. This value is used for performing setup checks inside the current design.

When *-min* or *-max* qualifiers are not specified, the same specified value is considered for both qualifiers. Usually, the *-min/-max* qualifiers are not used, and the value corresponding to the max delay is specified.

For current technologies in the realms of nanometers, the hold values for individual flops have come down drastically, to the extent that many times these are negative. Obviously, because of some finite delay in the data path, the input arrival time would be positive and would meet the negative hold value. Even if the hold value is not negative, it would be very close to *0*. Thus, in most cases, the delay through the launching device (launching flop's *Clk_to_q* delay) and the interconnect delay will be more than the hold time. Thus, an externally arriving signal would meet the hold requirements in most cases. The value used is typically the max value – so that the setup check is performed reliably. The same value also gets used for hold checks, which anyways gets met. If the signal is going to feed into an element which has a large hold requirement, say a memory, the hold check could become important. In such a situation, the *-min* value also should be specified correctly.

9.3.5 *-add_delay*

Most of the timing analysis tools provide an interactive shell to the designers. These shells allow users to modify a previously specified value. If an input delay is specified for a port, the current specification overrides prior specifications of input delay on this port. If a user has to specify input delays with respect to multiple reference events on the same port, then *-add_delay* needs to be specified for all subsequent specifications. For the first delay specification on the input port, it is not necessary to specify *-add_delay*. However, even if a designer puts an *-add_delay* to the first specification, the option would simply be ignored. Presence of *-add_delay* switch tells the tool that this is an additional constraint, besides the existing constraints, and it does not override the previous constraints. In the absence of this switch, the previous constraints get overwritten, for example:

first specification
set_input_delay -clock CLK1 -min 3.0 [get_ports In1]

this is OK, since, -max was not specified on this port earlier
set_input_delay -clock CLK1 -max 4.0 [get_ports In1]

this overrides both the previous specifications, with respect to CLK1
set_input_delay -clock CLK2 3.5 [get_ports In1]

This one gets added, without removing any of the earlier existing constraints
set_input_delay -clock CLK1 -add_delay 3.0 [get_ports In1]

This is also OK, since, In2 did not have any input_delay specified on it earlier
set_input_delay -clock CLK3 4.0 [get_ports In2]

9.3 set_input_delay

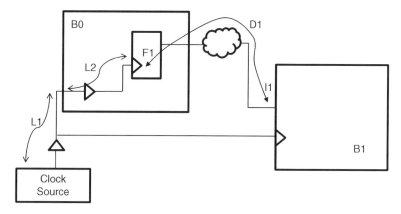

Fig. 9.11 Clock latency impact on input delay

9.3.6 Clock Latency

In Chap. 8, we saw that a clock suffers some delay (source latency and network latency) in its path, before the actual trigger reaches flops. Typically, the delay values at the inputs are specified with respect to the clock reaching the triggering flop. Consider the circuit shown in Fig. 9.11.

The delay value *D1* specified is from *F1*'s clock terminal till the *I1* pin of *B1*. The timing analysis tool will be able to consider the clock delay from the source till *F1*'s pin – based on the latency specified for the clocks. *L1* represents source delay and *L2* represents network latency. So the timing analysis tool considers the clock edge reaching *F1* at *L1+L2* after the active edge at the source. Then, it assumes that the signal will reach at *I1*, after a further delay of *D1* as mentioned in the *set_input_delay* specification.

However, if the designer has already included the source latency or the network latency, in the specified value, the switches -*source_latency_included* and -*network_latency_included* need to be mentioned. If *L1* has been included, then the switch -*source_latency_included* needs to be specified. If *L2* has been included, then the switch -*network_latency_included* should be specified. A user could specify none, either of or both of these switches, depending upon which portion of the clock path latency has been included.

When these switches are specified, the timing analysis tool will not consider the corresponding latency while computing the arrival time of the clock at the launch flop, thereby effectively causing the launch edge to be advanced.

These switches are usually not used because of the following reasons:

- Different tools treat these switches differently. For some tools or translator utilities, these switches are simply ignored. If you decide to use this switch, you need to make sure that all the tools in your flow treat these switches in a consistent manner.

- Both the launching device (outside the block of interest) and the capture device (inside the block of interest) will have similar latencies on the clock path. The timing analysis tool is anyways going to consider the clock within the block of interest. These switches only impact the clock edge on the launching side. It is always much easier and intuitive if both sides are treated similarly. Specifying these switches cause a different treatment on the launch side which is not very intuitive.
- It is much more intuitive to comprehend the delay from the launching device till the destination, rather than going further down the clock source also.
- From the perspective of reuse of constraints also, it is better to not include source and network latencies. Anytime, these latencies get changed; besides changing the latency related constraints, we would need to update the input delay specifications also.

9.3.7 Completing Input Delay Constraints

The only thing that is now left to complete the input constraints is to specify the actual ports or pins on which the input delay has to be specified and the delay values. These two are the only mandatory options to the *set_input_delay* command.

9.4 set_output_delay

The SDC command for specifying delays on output ports is *set_output_delay*. The BNF grammar for the command is:

```
set_output_delay    [-clock clock_name]
                    [-clock_fall]
                    [-level_sensitive]
                    [-rise]
                    [-fall]
                    [-max]
                    [-min]
                    [-add_delay]
                    [-network_latency_included]
                    [-source_latency_included]
                    delay_value port_pin_list
```

Conceptually, *set_input_delay* and *set_output_delay* commands are very similar. Instead of going through the explanation of each option, this section will only explain the differences (if any).

The fundamental difference between *set_input_delay* and *set_output_delay* is that input delay mentions the time that it takes to reach the input, which in turn means the time at which the signal would be available at the input. On the other hand, output delay mentions the time that the signal needs to travel after the output. This is different from the time at which the signal needs to be available at the output. However, fundamentally the semantics of both the commands are still similar in the sense that both these commands specify the delay requirements outside the block of interest. The block inside has to have its timings such that the external delays are met.

9.4.1 Clock Specification

-clock option is used to specify the reference clock that is used to sample the data coming out of the output port. Like *set_input_delay*, if the clock which samples the data does not enter the block of interest, we would need to specify a virtual clock with the same characteristics and use that virtual clock.

-clock_fall needs to be specified, if the capturing device is a falling-edge-triggered device.

Clock specification can be skipped, if the output pin is part of a combinational only path, though it is not very common to specify *set_output_delay* without the clock.

9.4.2 -level_sensitive

If the capturing device is a latch, rather than a flop, *-level_sensitive* can be specified. Conceptually, use of this switch allows for the condition that the output port could be borrowing the time from the capturing latch. That means the setup slack could increase – to account for signal reaching the latch after it has become transparent.

Like *set_input_delay*, this option should be used after a very careful consideration.

9.4.3 Rise/Fall Qualifiers

The *set_output_delay* command needs to provide the *-rise* or *-fall* qualifiers, if the output required times are different for a signal which is rising at the output port and a signal which is falling at the same output port.

9.4.4 Min/Max Qualifiers

-min and *-max* are used to qualify whether the delay value specified corresponds to the minimum or maximum time required for the signal to travel from the output port. The *-min* value is used for performing hold checks at the port/pin. The *-max* is used for performing setup checks at the port/pin.

Like *set_input_delay*, even for *set_output_delay*, usually min delays are not very important. And the max values are specified without min or max qualifiers.

9.4.5 -add_delay

Like *set_input_delay*, if a user has to specify output delay with respect to multiple reference events on the same port, then *-add_delay* needs to be specified for all subsequent specifications. Presence of *-add_delay* switch tells the tool that this is an additional constraint, besides the existing constraints, and it does not override the previous constraints. In the absence of this switch, the previous constraints on this port get overwritten.

9.4.6 Clock Latency

The output delay is considered with respect to the clock edge at the capturing flop's clock terminal. The timing analysis tool considers the clock edge reaching the capturing flop after the clock latencies. However, if the designer has already included the source latency or the network latency in the specified value, the switches *-source_latency_included* and *-network_latency_included* need to be mentioned.

When these switches are specified, the timing analysis tool will not consider the corresponding latency while computing the arrival time of the clock at the capturing flop, thereby effectively causing the capture edge to be advanced.

These switches are usually not used even with *set_output_delay*.

9.4.7 Completing Output Delay Constraints

The only thing that is now left to complete the output constraints is to specify the actual ports or pins on which the output delay has to be specified and the delay values. These two are the only mandatory options to the *set_output_delay* command.

9.5 Relationship Among Input and Output Delay

Let us consider the circuit shown in Fig. 9.12.

A signal goes from block *B1* to *B2*. For the ease of understanding, let us assume:

- The delay through *C1* includes the *Clk_to_Q* delay also.
- The delay through *C2* includes inter-block interconnect delay also.
- The delay through *C3* includes the setup requirement for the *F3* flop also.

9.5 Relationship Among Input and Output Delay

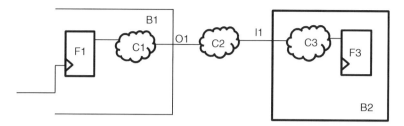

Fig. 9.12 Input and output delay relation

For this path, going from *F1* and into *F3*, an output delay needs to be specified on the output port *O1* for block *B1*, and an input delay needs to be specified on the input port *I1* for the block *B2*. We have already seen that the data takes one cycle to reach from one register to another. That means the total delay through *C1*, *C2*, and *C3* has to be less than or equal to one clock period (denoted by *P*). Thus,

$$C1 + C2 + C3 <= P \tag{9.1}$$

For the output at *O1*, the signal needs another *C2+C3* to be captured reliably at *F3*. So the *set_output_delay* (*SOD*) will be equal to or more than *C2+C3*.

$$SOD >= C2 + C3 \tag{9.2}$$

P-SOD denotes the time allowed for logic *C1*. Higher *SOD* means lesser time for logic *C1*.

For the input at *I1*, the signal takes *C1+C2* to reach *I1*. Thus, the *set_input_delay* (*SID*) will be equal to or more than *C1+C2*.

$$SID >= C1 + C2 \tag{9.3}$$

P-SID denotes the time allowed for logic *C3*. Higher *SID* means lesser time for logic *C3*.

The inequalities in (9.2) and (9.3) denote any additional margin/pessimism that the designer might put in. If the constraints are supposed to be just right, then all these inequalities can be converted to just an "equal to" sign. For the sake of ease, let us further assume that there is no additional margin and all constraints are applied to just meet the timing. In that case:

$SID+SOD = C1+C2+C2+C3$ (from (9.2) and (9.3)).
Or $SID+SOD = P + C2$ (from (9.1)).

If *SID+SOD* is higher than *P+C2*, it indicates that the design is over-constrained.

9.6 Example Timing Analysis

Let us do a sample timing analysis, based on our understanding so far. Depending upon which tool you use, your report would look different. The exact method of calculation could also be different. However, if you understand the basic calculation given in this section, it should be possible for you to understand any meaningful timing report.

For the sake of this example analysis, we will be making assumptions on various values. These assumptions will be explicitly mentioned, as we make those. We will use the same circuit as shown in Fig. 9.12.

9.6.1 Input Delay: Max

At flop *F3* in block *B2*:
Data arrival time:

Clock Rising Edge: Incremental: 0.0; Total 0.0 ## Assuming, active edge of reference clock is at 0.0; and rising edge is the active edge.
Clock Latency: Incremental: 0.5; Total 0.5 ## Assuming, clock source latency = 0.5
Clock Network Latency: Incremental: 0.7; Total 1.2 ## Assuming, clock network latency = 0.7
Input Delay: Incremental: 6.0; Total 7.2 ## Assuming, max input delay is specified as 6.0
Delay through C3: Incremental: 1.2; Total 8.4 ## Assuming, max delay through C3=1.2
Data available at F3's D pin: 8.4

Data required time:

Clock Rising Edge: Incremental: 10.0; Total 10.0 ## Assuming, next positive edge of clock triggering the flop is at time 10
Clock Latency: Incremental: 0.5; Total 10.5
Clock Network Latency: Incremental: 0.7; Total 11.2
Setup requirement: Incremental: 0.3; Total 10.9 ## Assuming, setup is 0.3.
Clock Uncertainty: Incremental: 0.2; Total 10.7 ## Assuming, clock uncertainty = 0.2
Data required at: 10.7
Slack = 10.7 − 8.4 = 2.3

Since the slack is positive, it implies that the setup timing on this path has been met.

Note:

- Incremental time denotes additional delay introduced at this stage
- Total time denotes cumulative delay till this stage. This equals incremental time of this stage + total time of the previous stage.

9.6 Example Timing Analysis

- Setup means data required time will be earlier than clock edge. So the setup value is subtracted, rather than added.
- Uncertainty is supposed to reduce the slack. It can either be added to the data available time or subtracted from the data required time. Either way, it reduces the slack. Many users typically say only the word "slack" when they mean "negative slack." In reality, positive slack means the timing is met and it is good. The negative slack is the one which causes trouble and needs fixing. Thus, higher degree of uncertainty is bad – as the slack gets reduced by that much.
- Usually, the delay through *C3* would be completely enumerated, explicitly showing all the elements and nets in the path and showing the delays through each of these elements.
- Clock latency values are used, when the clock network delays are not propagated and are used based on latency specifications. If the clock network delays are propagated, the actual clock path is also enumerated along with showing the delays through each of the elements in the path.

9.6.2 Input Delay: Min

At flop *F3* in block *B2*:
Data arrival time:

Clock Rising Edge: Incremental: 0.0; Total 0.0 ## Assuming, active edge of reference clock is at 0.0; and rising edge is the active edge
Clock Latency: Incremental: 0.5; Total 0.5 ## Assuming, clock source latency = 0.5
Clock Network Latency: Incremental: 0.7; Total 1.2 ## Assuming, clock network latency = 0.7
Input Delay: Incremental: 6.0; Total 7.2 ## Assuming, same value specified as min and max delays
Delay through C3: Incremental: 0.9; Total 8.1 ## Assuming, min delay through C3=0.9
Data available at F3's D pin: 8.1

Data required time:

Clock Rising Edge: Incremental: 0.0; Total 0.0
Clock Latency: Incremental: 0.5; Total 0.5
Clock Network Latency: Incremental: 0.7; Total 1.2
Hold requirement: Incremental: 0.3; Total 1.5 ## Assuming, hold is 0.3.
Clock Uncertainty: Incremental: 0.2; Total 1.7 ## Assuming, clock uncertainty = 0.2
Data required at: 1.7
Slack = 8.1 – 1.7 = 6.4

The slack is positive, which implies the hold timing on this path has been met.

Note:

- For the same logic *C3*, the delays considered for setup computations are higher than the delays considered for hold computation. This is because setup uses maximum delay through the path, while hold uses minimum delay through the path.
- Setup uses the next edge of the clock, because the requirement is for the data to be available before the next edge, while hold uses the current edge of the clock, because the requirement is for the data to not interfere with the data being sampled at the current edge.
- Hold means data required time will be later than clock edge. Hence, the hold value is added, unlike setup, which has to be subtracted.
- For setup checks, the requirement is that the data has to be available before the requirement. So slack=data required – data available. For the hold check, the requirement is that the data has to be available after the requirement. Hence, slack=data available – data required. The equation for slack computation is different for setup and hold.
- Uncertainty is supposed to reduce the slack. It can either be subtracted from the data available time or added to the data required time. Either way, it reduces the slack. The treatment (addition or subtraction) of uncertainty is different from setup, because the final slack equations are different for setup and hold. In either case, uncertainty is used to reduce the slack.

9.6.3 Output Delay: Max

At port *O1* of block *B1*:
Data arrival time:

Clock Rising Edge: Incremental: 0.0; Total 0.0 ## Assuming, active edge of clock triggering the flop is 0.0; and rising edge is the active edge.
Clock Latency: Incremental: 0.5; Total 0.5 ## Assuming, clock source latency = 0.5
Clock Network Latency: Incremental: 0.7; Total 1.2 ## Assuming, clock network latency = 0.7
Delay through C1: Incremental: 1.5; Total 2.7 ## Assuming, max delay through C1=1.5
Data available at O1: 2.7

Data required time:

Clock Rising Edge: Incremental: 10.0; Total 10.0 ## Assuming, next positive edge of reference clock is at time 10
Clock Latency: Incremental: 0.5; Total 10.5
Clock Network Latency: Incremental: 0.7; Total 11.2
Setup requirement: Incremental: 6.0; Total 5.2 ## Assuming, max output delay is specified as 6.0.

9.6 Example Timing Analysis   113

Clock Uncertainty: Incremental: 0.2; Total 5.0 ## Assuming, clock uncertainty = 0.2
Data required at: 5.0
Slack = 5.0 – 2.7 = 2.3

Since the slack is positive, hence, the setup timing on this path has been met.

Note:

- Setup means data required time will be earlier than clock edge. Hence, the setup value is subtracted, rather than added – similar to *set_input_delay*.
- The output delay specified is treated as a setup requirement (on a hypothetical flop which is driven directly by the output port).

9.6.4 Output Delay: Min

At port *O1* of block *B1*:
Data arrival time:

Clock Rising Edge: Incremental: 0.0; Total 0.0 ## Assuming, active edge of clock triggering the flop is at 0.0; and rising edge is the active edge
Clock Latency: Incremental: 0.5; Total 0.5 ## Assuming, clock source latency = 0.5
Clock Network Latency: Incremental: 0.7; Total 1.2 ## Assuming, clock network latency = 0.7
Delay through C1: Incremental: 0.5; Total 1.7 ## Assuming, min delay through C1=0.5
Data available at O1: 1.7

Data required time:

Clock Rising Edge: Incremental: 0.0; Total 0.0
Clock Latency: Incremental: 0.5; Total 0.5
Clock Network Latency: Incremental: 0.7; Total 1.2
Hold requirement: Incremental: -6.0; Total -4.8 ## Assuming, same output_delay value (6.0) is specified for min and max
Clock Uncertainty: Incremental: 0.2; Total -4.6 ## Assuming, clock uncertainty = 0.2
Data required at: -4.6
Slack = 1.7 – (-4.6) = 6.3

Since the slack is positive, hence, the hold timing on this path has been met.

Note:

- The output delay specified is treated as a hold requirement on a hypothetical flop, driven by the output port.
- The hold requirement is negative of the min *output_delay* specified.

9.7 Negative Delays

Sometimes, certain delay specifications can be negative.

set_input_delay -*min* being negative means the signal can possibly reach this input, even before the clock edge. This creates a need to have at least some delay for the input signal within the block, before it gets registered, else it might have a hold time violation on the flop trying to register it.

set_input_delay -*max* being negative means the signal will surely reach this input, even before the clock edge. This allows the signal to take more than the clock period within the block before the first register.

Input delays can be negative, if:

- There is minimal logic and interconnect that the signal has traveled outside this block (including the block which generated it and the top level routing).
- There is possibly a skew in the clock network, and the launching clock had reached earlier.

set_output_delay -*min* being negative means the signal can possibly take negative time outside the block, before it gets registered. In Sect. 9.6.4, we have seen that the hold requirement is negative of the min *output_delay* specified. So, for a negative value specified, the hold requirement becomes positive. This creates a need to have at least some delay within the block for the output signal.

set_output_delay -*max* being negative means the signal will surely take negative time outside the block, before it gets registered. So this allows the signal to take more than the clock period within the block, through the combinational logic between the last register and the output port.

Output delays can be negative, if:

- There is minimal logic and interconnect that the signal has to travel outside this block (including the block which is going to capture it and the top level routing).
- There is possibly a skew in the clock network, and the capturing clock will reach later.

Thus, negative value for min delay makes the hold requirement stringent, while negative value for max delay makes the setup requirement easier. Because, the delays through the devices and interconnects are typically more than the skew on the clock network, it is almost impossible to find a negative max delay. Min delays may sometimes be negative (very small delay through devices and interconnect), though it is not very common.

9.8 Conclusion

set_input_delay and *set_output_delay* need to be specified correctly, so that the signals traveling across the blocks get timed correctly. If this is not done correctly, individual blocks could meet their timing, but when the whole design is integrated,

9.8 Conclusion

the final design might not meet the timing, causing a lengthy debug and respin of affected blocks. Two of the most common mistakes that new designers make while specifying port constraints are:

1. While specifying set_input_delay, they use the sampling clock as the reference clock, while it should be the launching clock.
2. While specifying set_output_delay, they use the launching clock as the reference clock, while it should be the sampling clock.

In many designs, the sampling and the launching clock might be the same; hence, it might not matter. But, understanding this conceptual difference becomes important, if the two clocks are not the same.

Chapter 10
Completing Port Constraints

The delay through a specific cell depends on the slew/transition rate at its input as well as the load that it sees at its output. For cells inside the design, the fanin driver and the fanout cone is also part of the design. So the transition rate as well as the load can be computed by the tool. However, for the cells which are being driven by the input port, the input transition time is not known. Similarly, for the cells which drive the output port, the load is not known. Thus, designers need to provide the input transition time for the input signals and the external load that the output port will see. If not specified, the transition time is assumed to be *0*, namely, a sharp ramp (equivalent to infinite drive strength), and load is assumed to be *0*, namely, no external load. Both these conditions are highly optimistic.

The transition information can be specified through either of the following SDC commands:

set_drive
set_driving_cell
set_input_transition

The load information can be specified through either of the following SDC commands:

set_fanout_load
set_load
set_port_fanout_number

10.1 Drive Strength

Let us consider the circuit shown in Fig. 10.1.

Let us say the input *I1* of the block *B1* is being driven by the inverter. When the signal is transitioning to a *1*, the p-transistor of the inverter is driving the value on to the signal. This p-transistor offers some resistance to the driving signal.

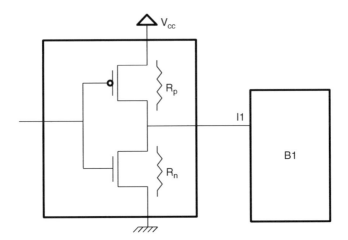

Fig. 10.1 Equivalent resistance

This resistance will influence the risetime at $I1$. When the signal is transitioning to a 0, the n-transistor's equivalent resistor influences the fall time at $I1$.

The resistive equivalent for the p-transistor and the n-transistor might not be equal. Thus, the driver-resistance value might be different for the input transitioning to a 1 or to a 0.

Now, let us say that the input is driven by a *NAND* gate, as shown in Fig. 10.2.

When $I1$ is transitioning to a 1, it could be because either $P1$ transistor is ON, or $P2$ transistor is ON, or both $P1$ and $P2$ transistors are ON. Depending upon which of the above situation is true, the resistive equivalent would be different. When both the transistors are ON, the resistance is minimal. Thus, the driver resistance could be within a range, rather than a specific value.

10.1.1 set_drive

The SDC command to specify the equivalent resistance of the driver is:

set_drive   [*-rise*] [*-fall*]
         [*-min*] [*-max*]
         resistance_value port_list

It should be noted that the value provided is actually the resistive value – which is inverse of the drive strength. Higher resistance means lower drive (longer time to transition) and vice versa.

-rise or *-fall* is used to specify whether the drive (actually, resistance of the driver) is for the signal rising or falling. When neither *-rise* nor *-fall* is specified, the specified value is considered to be applicable for both rising input and falling input.

10.2 Driving Cell

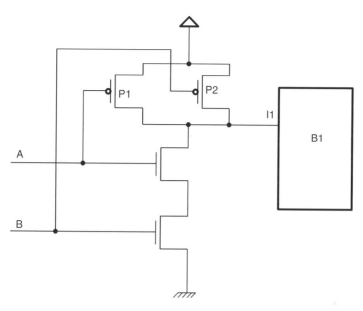

Fig. 10.2 NAND driver

-min or *-max* is used to specify whether the resistive value specified is the minimal resistance or the maximum resistance. Since minimal resistance means higher drive, hence faster transition, so minimal resistance is used for hold analysis. Similarly, max resistance is used for setup analysis. If none of *-min* or *-max* qualifiers are used, the specified value is applicable for both setup and hold analysis.

Given the resistive value of the driver, the tools can compute the slew time at the input, if they know the capacitive value of the input pin. It is worth reiterating that though the command is *set_drive*, the value specified is for resistance (which is inverse of drive).

Usually, *set_drive* is one of the less popular methods of specifying input slew.

10.2 Driving Cell

Most of the times, the internals of a cell are known only to the circuit designers for the ASIC library, rather than to the people who are using the libraries to create their designs. So it is much easier to specify the cell which will drive the input, rather than knowing the actual resistive value. The timing analysis tools have enough information from the library about the cell's electrical characteristics that they will be able to extract the relevant information.

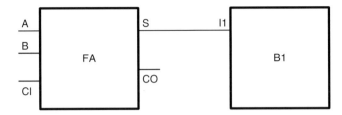

Fig. 10.3 Driver cell

10.2.1 set_driving_cell

The SDC command for specifying the driving cell is:

set_driving_cell [-lib_cell lib_cell_name]
                  [-rise] [-fall]
                  [-min] [-max]
                  [-library lib_name]
                  [-pin pin_name]
                  [-from_pin from_pin_name]
                  [-multiply_by factor]
                  [-dont_scale]
                  [-no_design_rule]
                  [-clock clock_name]
                  [-clock_fall]
                  [-input_transition_rise rise_time]
                  [-input_transition_fall fall_time]
                  port_list

10.2.1.1 Driver Cell Name

The -lib_cell switch is used to specify the cell which acts as the driver for the pin.

In Fig. 10.3, the input I1 of the block B1 is being driven by the cell FA. Thus, this cell should be specified as the -lib_cell.

Even though, SDC shows -lib_cell as an optional input, this switch is always found in any set_driving_cell command. Without this switch, the actual driver cell is not known. The rest of the command, options, or qualifiers might not have any meaning, if the driver cell itself is not known.

10.2.1.2 Min/Max, Rise/Fall

The -rise/-fall qualifier is used, when a designer wants to specify a different driving cell for a rise transition at the input pin and another cell for a fall transition at the input pin. If the qualifier is not used, then the same driver cell is used for both the transitions. Usually, a pin is driven by the same driver cell, irrespective of whether it is transitioning to a *1* or to a *0*. These qualifiers are needed, when the input is being driven by *pull-up* or *pull-down* kind of drivers. The *pull-up* cell can be specified with *-rise* and the *pull-down* cell can be specified with *-fall*.

The *-min/-max* qualifier is used, when a designer wants to specify different driving cells for setup (max) analysis and another driving cell for hold (min) analysis. Usually, the circuit would have the same driver; hence, it might appear surprising that different cells can be specified for setup and hold analysis. However, earlier in the design cycle, a designer might not know exactly which cell will drive this pin. So a designer might want to specify the strongest of the possible set of driver cells with *-min* option and the weakest of the possible set of driver cells with *-max* option. We've already discussed that hold analysis is given much less importance; thus, most often, the driver corresponding to the setup (max) analysis is specified.

10.2.1.3 Library

Sometimes, multiple libraries might be loaded in the tool. And there might be cells with the same name in more than one library. If the specified driving cell is found in multiple libraries, the tool might use its own mechanism to decide which of these cells should be considered as the driver. The switch *-library* is used to explicitly state the library from which the driver cell should be looked up. If only one library is loaded into the tool, or if the specified driver cell name exists in only one of the loaded libraries, this switch is not needed.

Generally, multiple cells with the same name are not simultaneously loaded into a tool. Even if the cells have same functionality, but if there is some difference in their electrical parameters, they are given a different name, e.g., *AN2* (2 input *AND* gate), *AN2H* (high-drive version of *AN2*), and *AN2LP* (low-power version of *AN2*). Thus, name clash usually does not happen for library cells. Some designers specify *-library* switch anyways – just to be explicit and ensure that only the intended cell gets specified.

10.2.1.4 Pin

For the example driver cell shown in Fig. 10.3, the driving cell has multiple outputs. Thus, the designer should specify *-pin* switch to clarify as to which of the outputs is being used to drive the input. For the circuit shown in Fig. 10.3, even if the driver (for *I1* pin of block *B1*) is known to be the *FA* cell, the drive strength of its output pins *S* and *CO* would be different. So the designer should mention the pin which drives the input (namely, pin *S* for the example given).

10.2.1.5 Timing Arc

A given output pin (either a cell with single output pin or for a multi-output cell – the pin specified with *-pin* switch) would have multiple timing arcs.

Fig. 10.4 Driver with multiple loads

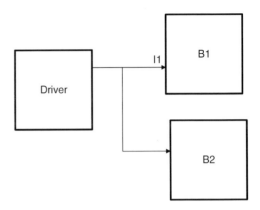

And the transition time on the output pin would depend on the timing arc that is chosen. *-from_pin* switch is used to specify the input pin of the driver cell, from which the arc should be chosen. So if the designer wants to specify that even for the *S* pin of the driver cell *FA*, the arc chosen should be the *A to S* path, then the *-from_pin* should specify *A*.

The switch may be useful when a designer wants to choose an extreme condition of timing and might want to specify the arc. Typically, if *S* pin is driving the input *I1*, then all transitions on *S* will reach the input *I1*.

10.2.1.6 Multiplication

The number specified with this option is the factor by which the computed transition time gets multiplied. Effectively, it specifies the factor by which the drive strength is considered to be reduced for the driver cell. Let us consider the circuit shown in Fig. 10.4.

For the input pin *I1* of the block *B1*, the driver cell can be specified. If needed, other option *(-library, -pin, -from_pin)* may also be specified in order to more accurately control the specific arc. However, the entire drive strength is not used to drive this input pin *(I1)*. The drive strength is used to drive another pin (on *B2*). So the effective drive strength applicable for *I1* pin of *B1* has to be divided. This effect can be achieved through the switch *-multiply_by*.

However, this effect is usually achieved through *set_load* command – as we will see later in Sect. 10.6.5 of this chapter.

10.2.1.7 Scaling

Certain electrical characteristics might need to be scaled based on the operating conditions. Another word for this scaling is *derating*. If you don't want the characteristics to be scaled or derated, you can use the switch *-dont_scale*.

10.2.1.8 Design Rules

If the driver pin has some design rule properties (e.g., the highest load that it can drive), those properties get transferred to the input port. The switch *-no_design_rule* prevents the properties from getting transferred to the input port. Any block boundary is only for use in modeling. In the realized circuit, all hierarchies would finally get dissolved. The signals being driven by the input port would finally get driven by the driver pin. Thus, all design rules applicable for the driver pin should be honored by the input port also, i.e., the design rules for the driver pin should get transferred on the input port.

10.2.1.9 Clock Association

If you want the driving cell to be specified only with respect to those *set_input_delay* which is specified with respect to a specific clock, the clock association can also be specified. Let us consider the following examples:

set_input_delay -clock CLK1 3.0 [get_ports I1]
set_input_delay -clock CLK2 4.2 [get_ports I1] -add_delay

Now, if we specify:

set_driving_cell BUF1 [get_ports I1]

it would mean that *BUF1* will be used as a driver for both the above input_delay specifications.

However, if we were to specify:

set_driving_cell BUF1 -clock CLK1 [get_ports I1]

that would mean that the driving cell, *BUF1*, would be used only for the first input_delay specification.

So the clock association can be used to limit the driving cell specification only for certain input_delay (those associated with the same clock).

Even if an input has delays specified with respect to multiple clocks, the pin would still be finally driven by a single cell. However, sometimes this option is used in conjunction with a few other options. Let us consider the circuit shown in Fig. 10.5.

For the input pin *I1* for the block *B1*, *set_input_delay* would be specified with respect to clocks *CLK1* and *CLK2*. A very accurate modeling of driving cell can be done using the following set of commands:

set_driving_cell -lib_cell MUX21 -from_pin A -clock CLK1 [get_ports I1]
set_driving_cell -lib_cell MUX21 -from_pin B -clock CLK2 [get_ports I1]

Thus, for the paths originating from *CLK1*, the *A to Z* arc (of *MUX21*) gets used for the driving cell, and for the paths originating from *CLK2*, the *B to Z* arc gets used for the driving cell.

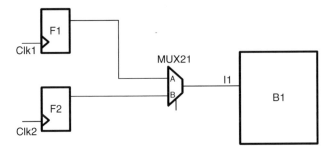

Fig. 10.5 Clock specification for driving cell

This is an example where even though some options individually do not make much sense, however, in combination with some other options, they are able to provide a much more accurate modeling.

10.2.1.10 Input Transition

These options are used to specify the rise and fall transition times at the input of the timing arc of the driver cell. Looking again at the circuit in Fig. 10.5, we could use -*input_transition_rise* and -*input_transition_fall* to specify the rise and fall transition values at the inputs *A* or *B* for the driver cell, *MUX21*. This rise and fall transition in turn will impact the transition time at the output *Z* of the *MUX21*, which will be seen at the input of the block *B1*.

It should be noticed that these transition times are not at the input of the block under consideration. Rather, these are the transition times at the input of the driving cell, which itself lies outside the block of interest. Clearly, this is second order of accuracy.

10.2.1.11 Ports

The designer has to specify the list of ports for which the driving cell property is being applied.

The most commonly used options for the *set_driving_cell* command are the name of the driving cell and the port for which the driving cell is specified. All other switches and options are used much less frequently. These other options are used under some specific situations, in order to achieve much finer control – as explained in above sections.

10.3 Input Transition

The commands *set_drive* and *set_driving_cell* are used by the tools to compute the transition time at the input port. However, a designer could specify the transition time directly. The SDC command for specifying the input transition time directly is:

set_input_transition   [*-rise*] [*-fall*]
                          [*-min*] [*-max*]
                          [*-clock* clock_name]
                          [*-clock_fall*]
                          transition port_list

The significance of *-rise*, *-fall*, *-min*, and *-max* qualifiers and the clock specifications are the same as mentioned for *set_driving_cell*. They are not being explained in this section to avoid repetition. The fundamental difference with respect to *set_driving_cell* is that usually, the driver cell remains unchanged, and hence, these qualifiers are not used that often with *set_driving_cell*. However, transition times are different for rise and fall or for min and max. Hence, *set_input_transition* often uses these qualifiers to specify different transition times for rise and fall and for setup and hold analysis.

The transition values specified are the actual transition times for the specified ports.

10.3.1 Input Transition Versus Clock Transition

In Chap. 8, we had seen *set_clock_transition* command. The main difference between *set_clock_transition* and *set_input_transition* is that for the transition time specified with the *set_clock_transition*, the specified value is used for the entire clock network, rather than computed for different points on the network, while the transition time specified with the *set_input_transition* is used only for the specified port. For all other points in the fanout cone, the transition time is computed. Let us consider the circuit shown in Fig. 10.6.

If the transition is specified on *clk* port using *set_clock_transition*, then the same transition value will be used at the clock terminals of all the flops in the fanout cone of the *clk* port. For all the nets in the network, namely, *n1*, *n2*, *n3*, …, the same transition time would be assumed to be applicable.

However, if the transition time is specified on *clk* port using *set_input_transition*, then the transition rate would be computed at each of the points, namely, *n1*, *n2*, *n3*, … – including the clock terminals of all the flops.

A good usage of these two constructs is: Before the clock tree is routed, the clock net could be driving a huge fanout. Trying to compute the transition time on the clock net could result in a very very slow transitioning signal – due to a heavy load. So, in such cases, *set_clock_transition* should be used. It is expected that the clock tree synthesis will ensure that the transition rate on the clock network remains

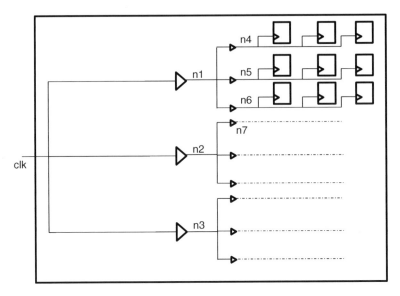

Fig. 10.6 Clock network

within a reasonable value. Once the clock tree has been synthesized, it is better to actually compute the transition value at all points. Use of *set_input_transition* (instead of *set_clock_transition*) at this stage will cause the timing analysis to compute the real transition value in the fanout cone of the *clk* port.

10.4 Fanout Number

Many wire-load models depend on the number of fanout pins to estimate the effective wire capacitance. The SDC command to specify the fanout number is:

set_port_fanout_number value port_list

This is a very simple command that specifies the fanout count of various output ports. This command has no implication, if external parasitic load is known and is being specified (through *set_load* command explained in Sect. 10.6). Since the value is the number of pins in the fanout of the port, it is expected to be an integer.

10.5 Fanout Load

Let us consider the circuit shown in Fig. 10.7.

The output *O1* drives two pins, a buffer and an *AND* gate. However, the load exhibited by the buffer is different than the load exhibited by the *AND* gate. The fanout number as mentioned in the previous section just considers the number

Fig. 10.7 Fanout load

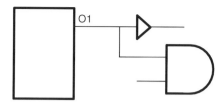

of pins being driven. It does not give any idea of the effective loading that those pins together exhibit. The output port needs to know the total effective load that it sees. SDC allows designers to specify the total load in terms of multiple of standard fanout loads. The standard load is defined in the library. The SDC command to specify the external load in terms of standard load is:

set_fanout_load value port_list

Like *set_port_fanout_number*, this is also a simple command. Let us consider that the load exhibited by the buffer is one standard load. And also assume that the load of the *AND* gate is *1.5* times the load of the buffer. So AND gate's load becomes *1.5* standard load. Thus, the load seen by the port *O1* is *2.5* (standard load). This example should also explain the fundamental difference between *set_fanout_load* and *set_port_fanout_number*. For the same circuit, the *set_port_fanout_number* would be *2*, because the port drives *2* pins.

10.6 Load

When external load is expressed in terms of standard load, the tools convert the fanout values into the equivalent capacitive load. A more commonly used style of expressing external load is by specifying the external capacitance value directly, rather than the fanout. The SDC syntax for *set_load* command is:

set_load   [*-min*] [*-max*]
           [*-subtract_pin_load*]
           [*-pin_load*]
           [*-wire_load*]
           value objects

The *-min/-max* qualifiers have the usual meaning. The *-min* value is to be used during hold analysis, and the *-max* value is to be used during setup analysis.

10.6.1 Net Capacitance

One of the most important things to note is that *set_load* can be applied even on wires, which are internal to the design under analysis. Thus, it provides a very convenient method to annotate extracted net capacitance obtained after post-layout.

Fig. 10.8 Pin load adjustments for net capacitance

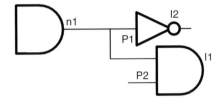

After the layout is done, effective net capacitance of each net can be extracted. And for the timing analysis tools, the capacitance value can be annotated on each net, through the *set_load* command. This allows net capacitance to be extracted by tools which are more accurate in extraction and used by STA tools.

10.6.2 Pin Load Adjustments

Let us consider the circuit shown in Fig. 10.8.

Let us say that during extraction of net capacitance for net *n1*, the extraction tool also included the loading of the pins *I1/P2* and *I2/P1*. Now, this capacitive value gets annotated on the net *n1*. The timing analysis tool sees the load on net *n1* which it considers to be the wire load only and then adds the load due to pins, *I1/P2* and *I2/P1*. So the load due to the pins gets counted twice.

So, in order to avoid double-counting, the switch *-subtract_pin_load* needs to be specified while annotating net capacitance. However, this switch should be specified only if the pin capacitances were also included during extraction. Most extraction tools will extract the net capacitance separately. If only that net capacitance is being annotated, this switch should not be specified.

Designers need to understand their extraction methodology, before deciding whether or not the pin load adjustments have to be made when specifying load on nets.

10.6.3 Load Type

Whether the specified load is of type pin or wire is specified using switches *-pin_load* or *-wire_load*. Tools might treat wire and pin loads differently. For example, slew might be degraded when propagating through wires. Let us consider the circuit in Fig. 10.9.

The slew *s1* at the output of the *AND* gate is computed based on the characteristics of the *AND* gate, the slew at the corresponding input pin of the gate and the load at the gate output. This signal then moves across the wire to the input of the buffer. However, as the signal travels across the wire, the slew characteristics get changed by the time it reaches the input of the buffer, so that the slew *s2* at the buffer input is

Fig. 10.9 Slew degradation through wire

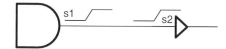

different from the original slew, *s1*. This concept called slew degradation occurs for wires and not for pins. Hence, it is important to convey whether the specified capacitive load is for net or for pin.

10.6.4 Load Versus Fanout Load

set_load and *set_fanout_load* differ in the sense that *set_load* specifies the actual capacitive value of the load, while *set_fanout_load* specifies the load in terms of a standard load.

Capacitive load (specified by *set_load*) = standard loads (specified by *set_fanout_load*) × capacitive load of single standard load.

During earlier days, standard load used to be more commonly used mechanism to specify pin loads; however, standard loads are not used that commonly in current technology libraries.

10.6.5 Load at Input

Usually, load is specified at output ports. However, sometimes load might need to be specified at input ports also. Let us revisit the circuit in Fig. 10.4.

For the block *B1*'s *I1* pin, if drive strength or driver is specified and the same driver is seeing additional load, then the effective drive available to *I1* pin is reduced. That additional load needs to be specified as a load on the input pin *I1* so that the effective drive can be adjusted accordingly.

This load needs to be specified only if the drive at *I1* is being specified as drive strength (*set_drive*) or a driving cell (*set_driving_cell*). If an input transition is specified for *I1*, the load at the input does not have any impact.

10.7 Conclusion

With the transition times and the load values specified, the inputs and outputs are fully constrained. The bidirects need to be constrained as if they are both inputs (i.e., input transition) and outputs (i.e., load). Usually, clocks or reset pins are driven by higher drive cells. Thus, drive strengths for clock/reset ports/pins are usually different from the drive strengths for other functional pins.

Though there are several different commands, the tools actually need input transition time and the output load. If the information is provided in another form (e.g., drive, driving cell, standard load), then the information is transformed into input transition and output load.

Typically, during the early stage of a design, the actual details of the driver are not known, as all the blocks are being built bottom-up; at this stage, it is better to constraint the input transition through commands like *set_input_transition*. As the design progresses and various modules and the top level SoC are synthesized and available, the actual driver cell is known. At this stage, the actual drivers and arcs can be specified using *set_driving_cell*.

For output specification, the more commonly used command is *set_load*.

Chapter 11
False Paths

11.1 Introduction

So far we saw how you can constrain your clocks and ports to specify the timing requirements for the design. However, even after setting these global requirements, designers would want to make certain exclusions for certain paths. This may be done to specify certain unique requirements on the paths or provide additional scope for leniency. Such constraints are referred to as timing exceptions. There are three kinds of timing exceptions:

1. False paths – These are paths that don't need to meet any timing requirements. Implementation tools ignore timing on such paths when constrained.
2. Multi cycle paths – These are paths that need more than one cycle to propagate data. Implementation tools relax timing on such paths.
3. Min and max delay – These are paths with specific maximum and minimum delay requirements and specified when designers want to override inferred setup and hold requirements. In this chapter, we will focus on false paths.

11.2 set_false_path

False path is modeled in SDC using *set_false_path* command.
The BNF grammar for the command is:

set_false_path   [-*setup*]
                      [-*hold*]
                      [-*rise*] [-*fall*]
                      [-*from* from_list]
                      [-*to* to_list]
                      [-*through* through_list]
                      [-*rise*_from rise_from_list]

[-*rise_to* rise_to_list]
[-*rise_through* rise_through_list]
[-*fall_from* fall_from_list]
[-*fall_to* fall_to_list]
[-*fall_through* fall_through_list]
[-*comment* comment_string]

Between all these switches, the command specifies:

- The exact path(s) which are to be treated as false
- The transitions within the paths which are to be treated as false
- Whether the false path relationship is for setup or for hold
- Any additional textual annotation to explain the context/justification for the false path

11.3 Path Specification

The paths which are to be declared as false path are identified using *-from*, *-through*, and *-to* options. The specification could include one or more of the above options. *-from* and *-to* can be specified at most only once each for each command. However, *-through* can be specified multiple times within the same command. Each of the *-from*, *-through*, and *-to* can take several points in the design as its argument. Let us consider a design represented as a graph network, as shown in Fig. 11.1.

Each point in the graph (i.e., *S1, S2, S3, S4, P1, P2, P3, P4, X1, X2, X3, X4, D1, D2, D3, D4*) represents a circuit element in the design. An edge in the graph

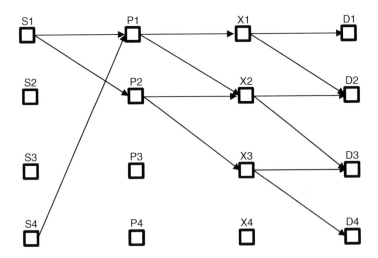

Fig. 11.1 A circuit represented as a graph network

11.3 Path Specification

represents a net connecting two elements. The following examples would show how these options are used to define the path specifications.

set_false_path -from S1 means all unique paths starting from *S1*. Thus, the path specification covers the following eight paths:

$S1 \rightarrow P1 \rightarrow X1 \rightarrow D1$
$S1 \rightarrow P1 \rightarrow X1 \rightarrow D2$
$S1 \rightarrow P1 \rightarrow X2 \rightarrow D2$
$S1 \rightarrow P1 \rightarrow X2 \rightarrow D3$
$S1 \rightarrow P2 \rightarrow X2 \rightarrow D2$
$S1 \rightarrow P2 \rightarrow X2 \rightarrow D3$
$S1 \rightarrow P2 \rightarrow X3 \rightarrow D3$
$S1 \rightarrow P2 \rightarrow X3 \rightarrow D4$

Similarly, *set_false_path -through P1* would cover all paths passing through *P1*, and *set_false_path -to D1* would cover all paths terminating at *D1*.

If multiple of these options are specified, the command applies to paths which satisfy each of the options. Example:

set_false_path -from S1 -through X1 covers only the following two paths:

$S1 \rightarrow P1 \rightarrow X1 \rightarrow D1$
$S1 \rightarrow P1 \rightarrow X1 \rightarrow D2$

Sometimes, the argument to any of the options is a list having multiple elements, rather than a single element. In that case, it is equivalent to having multiple commands each with one element only. Example:

set_false_path -from S1 -through {X1, X2} is equivalent to the following two command excerpts together:
set_false_path -from S1 -through X1
set_false_path -from S1 -through X2

Either we use the first specification or the lower two specifications; they would cover the following paths:

$S1 \rightarrow P1 \rightarrow X1 \rightarrow D1$
$S1 \rightarrow P1 \rightarrow X1 \rightarrow D2$
$S1 \rightarrow P1 \rightarrow X2 \rightarrow D2$
$S1 \rightarrow P1 \rightarrow X2 \rightarrow D3$
$S1 \rightarrow P2 \rightarrow X2 \rightarrow D2$
$S1 \rightarrow P2 \rightarrow X2 \rightarrow D3$

Stated alternately, when a list is specified as an argument to an option, it means any one of the elements in the list. Thus,

set_false_path -from S1 -through {X1, X2} means paths starting from *S1* and passing through either of (*X1* or *X2*).

When *-through* is specified multiple times, it indicates that each of the *-through* have to be satisfied independently. Thus, *set_false_path -through P1 -through X1* means paths which pass through both *P1* as well as *X1*. Thus, the command excerpt would cover:

$S1 \rightarrow P1 \rightarrow X1 \rightarrow D1$
$S1 \rightarrow P1 \rightarrow X1 \rightarrow D2$
$S4 \rightarrow P1 \rightarrow X1 \rightarrow D1$
$S4 \rightarrow P1 \rightarrow X1 \rightarrow D2$

In case of multiple *-through*, their order is also important. Thus, *-through P1 -through X1* is not the same as *-through X1 -through P1*. The order of *-through* specifies the order in which the paths must cover the circuit elements. Thus, *-through P1 -through X1* will cover the four paths enumerated above. On the other hand, *-through X1 -through P1* will not cover any path, because there is no path which goes first through *X1* and then through *P1*.

Thus, *set_false_path -from S1 -through {X1, X2}* is different from *set_false_path -from S1 -through X1 -through X2*. The *"-through {X1, X2}"* indicates a path passing through either *X1* or *X2*, while *"-through X1 -through X2"* indicates a path which goes through both *X1* and *X2* and also in the order $X1 \rightarrow X2$.

The options *-from* and *-to* can only be timing start and end points, respectively. Also, these options can specify clocks, in addition to circuit elements. When clock name is specified, it means all sequential elements triggered by the clock and all input/output ports whose input/output delay is specified with respect to the specified clock. In fact, specifying clocks in *-from* and *-to* provides an easy way to cover many sequential elements. Thus,

set_false_path -from CLK1 means all paths originating from:

- All sequential elements triggered by *CLK1*
- And all input ports constrained with respect to *CLK1*

In general, there are recommendations on what should be specified as a start or end point. The start point should be a clock, primary input or inout port, a sequential cell, clock pin of a sequential cell, or a pin on which input delay has been specified. Similarly the end point should be a clock, primary output or inout port, a sequential cell, data pin of a sequential cell, or a pin on which output delay has been specified. Some tools may allow certain other points to be also specified as start or end points, but using those points will minimize the reuse of such false paths in an implementation flow, as such exceptions may not be accepted universally by all tools.

It is apparent that the same set of paths can be specified in multiple ways. However, it is always advisable that the path specification should be done in a manner so that the intent is apparent from the command itself. Thus, if there is a false path requirement due to clock relationship, it is better to provide the specification in terms of clocks, rather than enumerating the individual flops.

11.4 Transition Specification

Sometimes, a designer might want to provide false specification for only a specific transition and not all transitions along the path.

This can be achieved through the usage of:

-rise_from: impacts only rising transitions at the specified start points
-fall_from: impacts only falling transitions at the specified start points
-rise_through: impacts only rising transitions at the specified through points
-fall_through: impacts only falling transitions at the specified through points
-rise_to: impacts only rising transitions at the specified end points
-fall_to: impacts only falling transitions at the specified end points
-rise: impacts only rising paths
-fall: impacts only falling paths

For most tools, the "rising path" or "falling path" is characterized in terms of the transitions at the end points. Thus, options *-rise* and *-fall* become synonyms of *-rise_to* and *-fall_to*.

These transition related options implicitly also specify a *-from* or *-through* or *-to*. Hence, a *-rise_from* should be seen as also having specified a *-from* implicitly, and *-from* should not be specified explicitly, if *-rise_from/-fall_from* is specified. Let us consider Fig. 11.2. In this figure, the clock network fans out to a network of flops, some of which are positive edge triggered and some of which are negative edge triggered. Further let us assume that all the paths in the combinational cloud are non-inverting.

set_false_path -from CLK will not time path from flops *F1* and *F2*.
set_false_path -from CLK can be written as combination of the following two command excerpts:
set_false_path -rise_from CLK
set_false_path -fall_from CLK

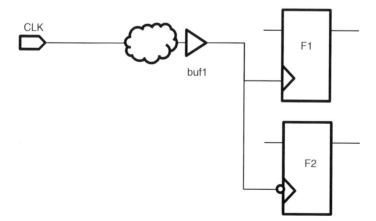

Fig. 11.2 Clock network with non-inverting paths but registers triggered by both edges

set_false_path -rise_from CLK will not time path from flop *F1*, while *set_false_path -fall_from CLK* will not time path from flop *F2*. *set_false_path -from CLK -fall_through buf1* does not impact the transitions which have a rising transition at *buf1*. Since all the paths in this design are non-inverting, this is equivalent to:

set_false_path -fall_from CLK -through buf1 which in turn is also equivalent to:
set_false_path -fall_from CLK -fall_through buf1

These set of equivalent commands will therefore result in path from flop *F2* not to be timed.

On the other hand, the following is an example command excerpts which will not impact any path:

set_false_path -fall_from CLK -rise_through buf1

Any rise transition at *CLK* can cause only a rising transition at *buf1*. So there is no path which qualifies the criterion of a falling transition at *CLK* and a rising transition at *buf1* simultaneously.

When a clock is specified as *-rise_from/-fall_from/-rise_to/-fall_to*, the specification implies the transition at the clock source (not at the elements triggered by the clock).

An application of this option could be the reset network. If the reset is asynchronous, there is no need to time the path for assertion of reset. Assuming this is negative-edge-triggered reset, this can be modeled as

set_false_path -fall_from reset_n

To understand the above command, let us consider the concept of recovery and removal timing check. Recovery timing check ensures there is sufficient time between an asynchronous signal going inactive and the next active clock edge. This setup like check therefore makes sure that the design has enough time to recover from reset before the next clock edge becomes effective. Similarly the removal timing check ensures that there is sufficient time between active clock edge and deassertion of an asynchronous control signal. This hold like check therefore makes sure that effect of asynchronous signal remains on the design and is not impacted by any active clock edge during that period. Figure 11.3 shows the waveform depicting these checks.

A good usage style of asynchronous reset recommends that while assertion can happen asynchronously, the deassertion should happen synchronously. This means the user would not like to time the path during assert, but all registers would need to come out of the reset state in the same cycle. Hence in this case, the user would need to ensure that timing checks are satisfied for deassertion. However, he doesn't care about the timing check during assertion. Assuming this is negative-edge-triggered reset, this can be modeled as

set_false_path -fall_from reset_n

If *-fall_from* option is not specified, then the false path gets applied for both assertion and deassertion of the asynchronous reset.

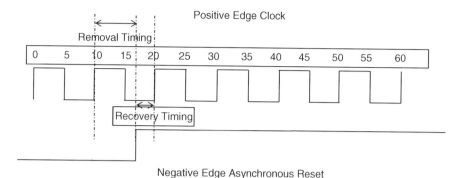

Fig. 11.3 Recovery and removal timing check

11.5 Setup/Hold Specification

In some situations the designer may want to set this constraint only for the setup or the hold path. This is achieved using the *-setup* and *-hold* options.

11.6 Types of False Paths

False paths are paths that may or may not exist in the design; however, timing on such paths doesn't make sense. In general, false paths are defined as paths where sensitization at the start point of the path causes no transition at the end point of the path. However, in some cases the path may exist and be possibly sensitized, but the user may not want to time the paths. The advantage of identifying false paths is that it gives guidance to implementation tool to only spend time optimizing real paths and meet timing on it. There are different kinds of false paths:

11.6.1 Combinational False Path

Let us consider the schematic in Fig. 11.4. In this figure path from $A \rightarrow mux1/A \rightarrow mux2/A \rightarrow B$ cannot be sensitized since the select signals of the two multiplexers have opposite sense. Such a false path, which cannot be sensitized and the control logic influencing the sensitization consists of combinational elements, is called a combinational false path.

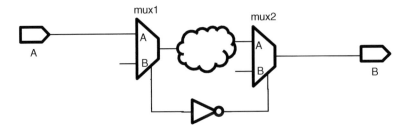

Fig. 11.4 Combinational false path

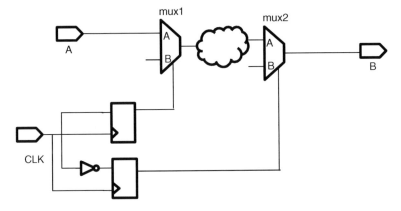

Fig. 11.5 Sequential false path

This false path can be modeled as:

set_false_path -from [get_ports A] -through [get_pins mux1/A] -through [get_pins mux2/A] -to [get_ports B]

In this example, the *-from* and *-to* need not be used. Any path through *mux1/A* and *mux2/A* should be false. By putting *-from A* and *-to B*, we are not covering other paths – which could be passing through this. This false path should be written as:

set_false_path -through [get_pins mux1/A] -through [get_pins mux2/A]

11.6.2 Sequential False Path

Let us consider the schematic in Fig. 11.5. This figure is a slight variant of Fig. 11.4, where in the select signal is driven by a one hot controller. In this figure as well, the path from $A \rightarrow mux1/A \rightarrow mux2/A \rightarrow B$ cannot be sensitized for the same reason. However, the control logic influencing the sensitization consists of sequential elements. Such a false path is called a sequential false path. The false path specification for such a path would also be the same as mentioned in the previous section.

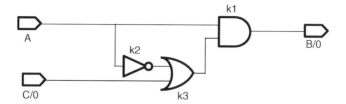

Fig. 11.6 Dynamically sensitized false path

11.6.3 Dynamically Sensitized False Path

Let us consider the schematic in Fig. 11.6. In this figure, there is a re-convergence of combinational path from A to B. There are two paths from A to B, namely, $A \rightarrow k1 \rightarrow B$ and $A \rightarrow k2 \rightarrow k3 \rightarrow B$. Now for the path $A \rightarrow k2 \rightarrow k3 \rightarrow B$ not to be blocked, the value of port C must be zero. If that be the case, then irrespective of the value on A (0 or 1), the value of B is always 0. So the path ($A \rightarrow B$) is never sensitized and appears to be a false path.

However, if the delay on paths $A \rightarrow k1 \rightarrow B$ and $A \rightarrow k2 \rightarrow k3 \rightarrow B$ is different, such that

delay on path ($A \rightarrow k1 \rightarrow B$) < delay on path ($A \rightarrow k2 \rightarrow k3 \rightarrow B$)

In this case a $0 \rightarrow 1$ transition on A will cause a glitch. And if

delay on path ($A \rightarrow k1 \rightarrow B$) > delay on path ($A \rightarrow k2 \rightarrow k3 \rightarrow B$)

In this case a $1 \rightarrow 0$ transition on A will cause a glitch. So irrespective of which path has higher delay, there will be a glitch – for one transition or the other breaking the inherent definition of a false path. Such a path where depending on the delay, the path may have a glitch is called a dynamically sensitized false path.

Such paths should not be declared as false. The glitch may happen just at the time that the signal is captured, causing the glitch to be captured, and thus, the design may fail.

11.6.4 Timing False Path

Timing false paths are paths that exist in the design and can be sensitized, but the designer chooses not to time them. For example, design contains configuration registers which get initialized during initialization sequence to bring the whole design to a known state for correct operation. After this, they maintain a static value and timing for these registers is not a designer care about.

Another example is path to asynchronous signals like reset. To initialize the FSMs in a design, the registers are reset so as to reach a known state before the first clock transition. Once the normal operation starts, such reset paths are not required to be timed.

Fig. 11.7 Protocol-based data exchange

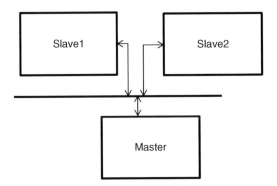

False paths are also used to model asynchronous domain crossings. We saw in Chap. 7 how *set_clock_groups* could be used to model this. Prior to SDC1.7, when *set_clock_groups* was not part of the standard, asynchronous clock domain crossings were modeled using false paths. Therefore, Fig. 7.1 could also be modeled as:

set_false_path -from C1 -to C2
set_false_path -from C2 -to C1

instead of

set_clock_groups -asynchronous -group C1 -group C2

As it can be seen, *set_clock_groups* is a more concise and efficient way of describing this relationship. More importantly, it conveys the intent correctly.

11.6.5 False Path Due to Bus Protocol

Let us consider a bus protocol-based design, as shown in Fig. 11.7.

Let the master and the two slaves be connected to the same bus. Further the protocol allows the master to initiate data transfer in either direction with either of the slaves. However, the slaves themselves might not be able to exchange data between themselves directly. Due to bidirectional connection between the bus to each of the slave peripherals, it might appear that there is a path between *Slave1* and *Slave2*, while, in reality, data will never get transmitted along this path. Thus, the path between *Slave1* and *Slave2* needs to be declared as false – in both directions.

Usually, these kinds of false paths are very difficult to determine or be verified by any tool, since the information is specific to the protocol and that is not apparent by looking at the design connectivity.

11.6 Types of False Paths

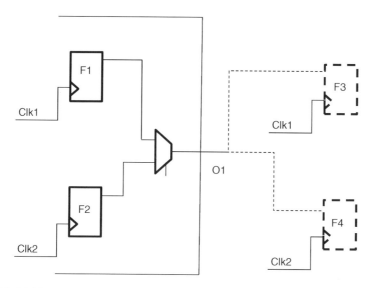

Fig. 11.8 Multiplexed output pin

11.6.6 *False Path Between Virtual and Real Clocks*

Today, more and more designs are being pad-limited, because the size of IOs is not reducing at the same pace as the circuit geometry. The size of the chip is being decided by the number of IOs, rather than by the amount of logic sitting inside it. This has resulted in more and more pins being multiplexed. Figure 11.8 shows one such multiplexed output pin.

Signals from *F1* and *F2* are multiplexed to use the same output pin *O1*. This can be done, if both *F1* and *F2* do not need to send outputs simultaneously. Let's further assume that when data is sent out by *F1* (clocked by *Clk1*), it is consumed outside the block by flop *F3*, which is also clocked by *Clk1*. Thus, an output delay has to be specified on *O1* with respect to *Clk1*. Similarly, let us assume that when the data is sent out by *F2* (clocked by *Clk2*), it is consumed by flop *F4* – clocked by *Clk2*. Hence, an additional output delay has to be specified on *O1* with respect to *Clk2*.

So *O1* has two output delays, one with respect to *Clk1* and another with respect to *Clk2*. And *O1* receives data from two sources: flop *F1* (triggered by *Clk1*) and flop *F2* (triggered by *Clk2*).

Thus, at *O1*, the following four checks are made:

1. Data starting from *F1* (triggered by *Clk1*) and output delay specified with respect to *Clk1*
2. Data starting from *F1* (triggered by *Clk1*) and output delay specified with respect to *Clk2*

3. Data starting from *F2* (triggered by *Clk2*) and output delay specified with respect to *Clk1*
4. Data starting from *F2* (triggered by *Clk2*) and output delay specified with respect to *Clk2*

Out of these four checks, only the (1) and (4) are of interest. The other two are not of interest, because data starting from *F1* is not supposed to be captured on *Clk2*, or data starting from *F2* is not supposed to be captured on *Clk1*.

Thus, these two checks, which are not of interest, need to be disabled. This can be done through *set_false_path* (or *set_clock_groups*) between *Clk1* and *Clk2*. This has the risk that any interaction between *Clk1* and *Clk2* elsewhere in the design also doesn't get timed.

Thus, a common practice is to create virtual clocks (say, *vClk1* and *vClk2*) corresponding to the clocks *Clk1* and *Clk2*, and the output delays are declared with respect to these virtual clocks. Now, false paths can be declared from the real clocks to the virtual clocks. Thus, the complete set of constraints would be:

create_clock -name vClk1 <period and waveform>
create_clock -name vClk2 <period and waveform>

set_output_delay <delay value> -clock vClk1 [get_ports O1]
set_output_delay <delay value> -clock vClk2 [get_ports O1] -add_delay

set_false_path -from Clk1 -to vClk2
set_false_path -from Clk2 -to vClk1

Similarly, for situations where an input pin is multiplexed to receive data from two different clocks, which are also sampled by the two corresponding clocks, the input delays should be specified with respect to the virtual clocks, and false paths can be declared between the virtual clock (used for specifying the input delays) and the real clock (which trigger the capturing flops).

11.7 set_disable_timing

Let us consider an excerpt of a design as shown in Fig. 11.9.

Here, if delay computation starts from or reaches point *B*, it will keep on adding delay, as it keeps making subsequent iterations over the loop $I1 \rightarrow n1 \rightarrow I2 \rightarrow n2 \rightarrow I1$, and never really converge. Most tools have a mechanism to detect such loops and

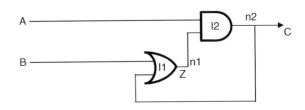

Fig. 11.9 Combinational loop

break the loop for the purpose of timing analysis. However, it is up to the tool to break whichever segment it considers appropriate. Let us consider that the tool broke the loop by breaking the segment *n1*. As a result, the path *B* to *C* also is not timed. In such situations, it is better for the user to specify the segment which should be broken for the timing analysis. The user needs to choose the segment in such a manner, so that no other paths get impacted. For example, in the current circuit, it is best to break the segment from *OR* gate's second input to its output. This breaks the loop as well as leaves all other paths intact for being timed.

Such breaking of the loop can be done through the command, *set_disable_timing*. The BNF grammar for *set_disable_timing* command is:

set_disable_timing   [*-from* pin_name]
  [*-to* pin_name]
  design_objects

Here the design object list includes cells, pins, and ports. For example, in the command shown below, the timing arc from *B* to *Z* for the *OR* gate (instance *I1*) would be removed from timing analysis:

set_disable_timing -from B -to Z I1

The conceptual difference between *set_false_path* and *set_disable_timing* is that former simply prevents timing of the paths; however, the delay calculation doesn't stop. However, in the case of *set_disable_timing*, the path itself is removed from timing analysis.

11.8 False Path Gotchas

While specifying the false paths, the designer must be careful about the following things:

1. Several tools allow users to use wildcards in the specification of false path. For example, in the circuit in Fig. 11.1 somebody could have tried to model as *set_false_path -from S**. This causes all paths from *S1, S2, S3, S4* to be made false. This can be very dangerous and can result in real paths from not being timed, which could result in timing failure on silicon.
2. Many times false paths are set between start and end points, which don't have a physical connectivity between them. Though harmless, they would result in longer and unnecessary runtimes in your implementation tools.
3. While specifying *-through*, the designer must ensure that the *through* is not redundant. When *-through* is specified, the design object on that list will not be optimized, even if it is a good candidate, which could result in suboptimal design implementation.
4. Many times different kinds of exceptions (false paths and multi cycle paths) are set on the same path or path segment. This kind of overlap causes implementation tools to make assumptions that may not be consistent.

11.9 Conclusion

In summary, false paths act as exceptions to timing constraints, which results in the path not to be timed. However, in some cases, design may require path to be timed but take more than one clock cycle to propagate the information. In the next chapter, we will learn how we can use multi cycle paths to provide scope for leniency beyond a single cycle.

Chapter 12
Multi Cycle Paths

By default, each path is timed for a single cycle, i.e., data launched at any edge of the clock should be captured by the next flop at the next rising edge of the clock on the destination flop. Figure 12.1 shows this relationship.

However, sometimes a designer might need to provide some additional cycles before the data is to be captured. Figure 12.2 shows this scenario of an additional cycle. The paths which get additional cycles are called multi cycle paths.

12.1 SDC Command for Multi Cycle Paths

The SDC command for declaring a path as multi cycle is:

set_multicycle_path   [*-setup*]
[*-hold*]
[*-rise*] [*-fall*]
[*-start*] [*-end*]
[*-from* from_list]
[*-to* to_list] [*-through* through_list]
[*-rise_from* rise_from_list]
[*-rise_to* rise_to_list]
[*-rise_through* rise_through_list]
[*-fall_from* fall_from_list]
[*-fall_to* fall_to_list]
[*-fall_through* fall_through_list]
path_multiplier
[*-comment* comment_string]

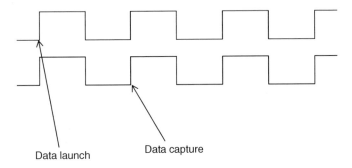

Fig. 12.1 Default setup timing relationship

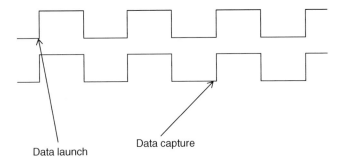

Fig. 12.2 Multi cycle of 2

Between all these switches, the command specifies:

- The exact path(s) which are to be treated as multi cycle
- The transitions within the paths which are to be treated as multi cycle
- Whether the multi cycle relationship is for setup or for hold
- Whether the additional cycle(s) are in terms of launch clock or capturing clock
- The number of cycles
- Any additional textual annotation to explain the context/justification for the multi cycle nature

12.2 Path and Transition Specification

Path and transition specification (options: -rise, -fall, -from, -to, -through, -rise_from, -rise_to, -rise_through, -fall_from, -fall_to, -fall_through) for a multi cycle specification is exactly same as that explained in Chap. 11. These options and path specifications are not being repeated here.

12.3 Setup/Hold Specification

For the purpose of this section, we will assume that both the launching device and the capturing device are triggered by the same clock. The implication of different clock frequencies will be discussed in Sect. 12.4. A clock waveform is depicted in Fig. 12.3.

Timing analysis will assume that the launching flop will launch the data at edge A. For setup analysis, it will consider that the data will be captured at the edge B. So, setup relation is analyzed between edges A and B. Use of -*setup* switch causes the capturing edge for setup analysis to be moved further to the right – away from A to C, D, etc., depending upon the number of cycles specified.

Let us assume that the capturing edge for setup has been moved to edge D, through -*setup* switch. For the purpose of hold analysis, the timing analysis tool considers the immediately preceding edge at the capture flop (when launch and capture clocks are the same). Thus, hold analysis will be considered using the edge C for capture. Use of -*hold* switch causes the capturing edge for hold analysis to be moved towards left, to either B or A – depending upon the number of cycles specified. The general practice is to restore the hold check back to edge A. If the hold check is not brought back to edge A, there might be buffers inserted in the path to ensure some delay. These buffers will take up silicon area as well as power.

We have assumed above, the setup edge was moved to 3 cycles (so that it reached edge D). The hold edge automatically moved to edge C (the immediately preceding edge). Now, in order to bring it back, it has to be moved back by 2 cycles. This can be achieved through use of -*hold* switch.

In order to move the hold check edge back to A, we have moved it by 2 cycles. It has now come back to the same edge as launch edge, i.e., at 0th edge (with launch edge being considered Origin). It should be noted that we are now talking about two different numbers. A hold multiplier number 2 which specifies the number of edges by which the hold edge needs to move towards left. This is the number that is specified in the *set_multicycle_path* with -hold. And, another is number 0, which specifies the actual edge number, where the check is happening. These two numbers are often a source of confusion during conversation. When you are talking about hold edge – specially in the context of multi cycle path – make sure that all the people involved understand, whether the number being mentioned is the "hold multiplier" the number by which the edge will move towards left, or the edge number, where the check will be performed. For the waveform shown in Fig. 12.3 (assuming setup number of 3), edge A corresponds to a hold multiplier number of 2, edge B corresponds to a hold multiplier number of 1, and edge C corresponds to a hold multiplier number of 0, in the context of the *set_multicycle_path* definition.

Fig. 12.3 Clock waveform

-setup switch specifies the period to which (not "by which") the setup capture edge will move to the right. Thus, a specification of *N* means move to *N*th period. This is different from move by *N* cycles. The *-hold* switch on the other hand specifies the period by which (not "to which") the hold capture edge will move to the left. If setup edge was moved to *N*th edge, the hold edge is automatically moved to *N–1*th edge. In order to restore it back to its original location, the hold check needs to be moved backwards by *N–1* cycles, so that it goes back to "*0*" edge.

Thus, *-setup* switch will move the capturing edge for setup. Simultaneously, it also moves the capturing edge for hold. After that, another *set_multicycle_path* might be needed with *-hold* switch to restore the hold check back to the original edges. Multi cycle path specifications are usually found in pairs of *-setup* and *-hold*. If the *-hold* specification is not given, the hold edge remains where it had got moved due to setup edge movement.

12.4 Shift Amount

The path multiplier specifies the number of clock cycles for the multi cycle path specification. If the launching device and the capturing devices are triggered by the same clock (or different clocks but with the same frequency), and the command specifies a multi cycle relationship, it is not important as to whether the number of cycles mentioned is for the start clock or for the end clock. However, if the start and the end clocks are different, then, it is important to specify whether the number of cycles specified is in terms of start clock or the end clock. For the purpose of our understanding, let us assume the clocks to be synchronous (Though the same approach can be extended to asynchronous clocks, however, asynchronous clocks are usually not timed).

Let start clock have a period of *10ns* and end clock have a period of *20ns*. For performing setup checks, the timing analysis tool identifies pairs of launch edge and the next capture edge. For all such pairs determined, the timing analysis tool figures out which of these pairs gives the minimal time for the data to travel. Figure 12.4 shows the waveforms for these clocks.

For the given example waveforms, the setup check would be made for the combination, launch at *B* and capture at *N*. So, for setup to be met, data path has to be within *10ns*.

Similarly, determine which are the worst case hold combinations, and use that launch/capture combination for hold check. The edge combinations used for hold check may not have any relation with the edges used for setup check.

For the given waveform, the hold check would be made for the combination: launch at *A* and capture at *M* (this is equivalent to launch at *C* and capture at *N*).

Within 1 cycle of destination clock, two data can get launched, which will mean losing one data. Within 1 cycle of destination clock, only one data should be launched – to avoid data loss. For a capture at *N*, the data could be launched either at *A* or at *B*. There is no advantage of launching at *B*, since the capture is still at *N*.

12.4 Shift Amount

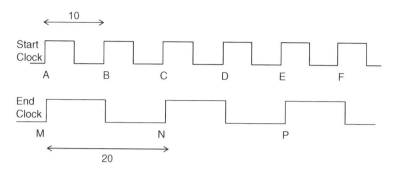

Fig. 12.4 Start and end clocks have different period

We might as well launch the data at A and provide 2 cycles (of source clock) for the data to reach the destination. Thus typically, for setup, we want to move the launch edge back to A. So, we declare a path of 2 cycles in terms of source clock. This can be specified through use of -*start* switch. This results in the launch edge being moved to the previous triggering edge of the launch clock, namely, launch at A and capture at N. So, setup gets another *10ns*. Note that the number "2" represents the number of cycles to be allowed for setup. It is different from the number of cycles by which the check got moved. The check got moved by 1 cycle, since the original setup check allowed for 1 cycle.

Now, for determining the hold check, the two pairs of edges are determined:

1. Launch at B and capture at N (hold launch edge one later than the setup launch edge).
2. Launch at A and capture at M (hold capture edge one earlier than the setup capture edge).

Out of these two combinations, the first one is more restrictive (higher requirement for hold check). Hence, the hold check moves to launch at B and capture at N. This means a minimum delay of *10ns* (B to N). The aim was to allow additional time, if needed, not to force a higher delay. So, we would want to restore the hold checks to default positions (viz., launch at C and capture at N). Thus, we need to move the launch edge towards right by 1 cycle of the start clock. This can be achieved by specifying -*start* switch with -*hold*. Now, the hold edges are C, N combination, which is same as A, M combination. Thus, the hold check edges have been restored to the original conditions. Note that in the case of hold multiplier, the number "1" represents the number of cycles by which the check got moved.

Let us consider another example. This time, the start clock has a period of *20ns* and the end clock has a period of *10ns*. Figure 12.5 shows the corresponding waveform:

For this combination of start and end clocks, the default setup check would be made at launch edge M and capture edge B. And, the hold check would be made at launch edge M and capture edge A.

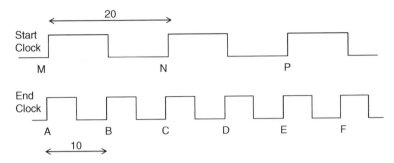

Fig. 12.5 Start clock slower than end clock

The data launched at *M* does not get changed for 2 cycles in terms of destination clock. So, multiple edges of destination clock would capture the same data (launched at *M*). We might as well disable the capture at all the edges of destination and do the capture only once for each launch. And, for a data launched at *M*, we would rather do the capture at *C*, thereby allowing more time for the signal to reach the destination. So, typically for setup, we declare a path of 2 cycles in terms of destination clock. This can be specified through use of -*end* switch. This results in the capture edge being moved to the next triggering edge of the destination clock, namely, launch at *M* and capture at *C*. Thus, setup gets another *10ns*.

This will cause the hold check to be moved to launch at *M* and capture at *B* (hold capture edge being 1 clock edge before the setup capture edge). If we want to restore the hold checks to default positions (viz., launch at *M* and capture at *A*), we need to move the capture edge towards left by 1 cycle of the end clock. This can be achieved by specifying -*end* switch with -*hold*.

So, effectively, for setup checks, -*end* means move the capture edge to the right, and -*start* means move the launch edge to the left. -*start* with -*hold* causes the launch edge to move to the right, and -*end* switch with -*hold* causes the capture edge to move to the left.

Stated alternately, -*start* moves the launch clock edge and -*end* moves the capture clock edge. With multi cycle, these edges move in a direction so as to make the checks less stringent. The number of cycles moved is always in terms of the edge that is moving. So, if launch edge is moving, the number of cycles is in terms of the launch clock.

Few more observations apparent from these two examples are:

1. In order to restore the hold edge back to the original location, the hold multiplier is 1 less than the setup multiplier.
2. For synchronous clocks with different frequencies, the setup number is equal to the ratio of the time period of the two clocks.
3. The multiplier is specified in terms of the period of the faster clock (smaller time period).

There should be an exact match in the number of data launched and data captured. For a one-to-one transmittal of data, that would mean one launch and one capture per one period of the slower clock. The remaining edges of the faster clock should be disabled, so that they neither capture nor launch additional data.

And the launch/capture edges can be so chosen that they provide maximum time for the data to travel. This will result in the observation 2 and 3 above.

These are just thumb rules. Each multi cycle path should be analyzed on its own for the right value of the multiplier, and the right clock period to be used.

The comment option can be used to specify a text annotation – mostly to mention the reasoning behind the multi cycle specification.

12.5 Example Multi Cycle Specification

Let us consider a few example situations of multi cycle specification. In the previous section, we already saw examples of synchronous data transfers between a fast to slow and a slow to fast clocks.

12.5.1 FSM-Based Data Transfer

Let us consider a circuit as shown in Fig. 12.6.

When the *data* is generated by *Cs*, the same clock also generates an *enable* signal. This signal goes through an FSM, and then the target capturing device is enabled to capture the data. Let us assume that the *enable* signal takes N cycles within the FSM, before the target device is ready to capture the data. In such a situation, there is no need for the actual *data* to rush to the target device immediately. It can take time up till N cycles (of destination clock) to reach there. Hence, this path needs to be constrained as:

set_multicycle_path -from Cs -through F1/Q -to Cd -setup N -end
set_multicycle_path -from Cs -through F1/Q -to Cd -hold N-1 -end

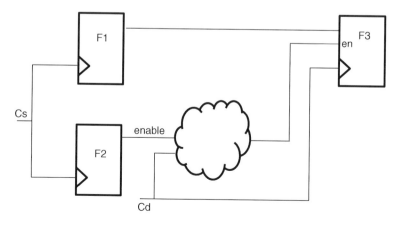

Fig. 12.6 FSM-based data transfer

Fig. 12.7 Simple realization of source synchronous interface

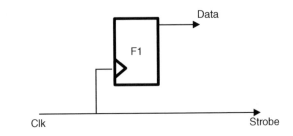

Fig. 12.8 Source synchronous interface – corresponding waveform

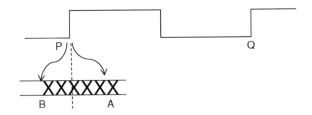

Note the use of *-through*. This is needed, so that only the *F1* to *F3* path are covered. The *enable* signal might need its own multi cycle specification, if it takes more than a cycle.

Paths using Handshake, Acknowledge, etc., for data transfer are examples of such situations which might use multi cycle. Another example is when a data bus is crossing a clock domain and instead of trying to synchronize the whole bus, an enable signal is synchronized. If the control FSM is going to take more than 1 cycle before enabling the capture device, the data can take those many cycles in reaching the destination device.

12.5.2 Source Synchronous Interface

In a source synchronous interface, as a data is presented at the output, the clock is also sent out. Both data and the clock lines can be routed on the board to have similar delays. Thus, the receiving device need not worry about the delays through the board trace. Whenever there is a clock signal, the receiving device knows that the actual data is also available around the same time. Figure 12.7 shows a typical realization of a source synchronous interface, and Fig. 12.8 shows the corresponding waveform.

In a system synchronous interface (in which all the signals are synchronized to system clocks), the output delay on the *Data* pin would be specified with respect to *Clk*. However, in the source synchronous interface, the reference is with respect to the *Strobe*. Thus, a timing relation specified wrt *Strobe* would cover *Data* pin also. The *Strobe* itself might be specified through *create_generated_clock* using *Clk* as the master.

12.5 Example Multi Cycle Specification

Usually, *Data* launched on *Clk* edge corresponding to edge *P* of *Strobe* would be timed for setup edge *Q*. However, in this interface, *Data* should be timed with respect to the edge *P* itself. Thus, the setup edge needs to be moved. This can be done through the command:

set_multicycle_path -from [get_clocks Clk] -to [get_clocks Strobe] -setup 0

Notice that the setup multiplier number is specified as *0*, which moves the setup edge towards left – to the same edge as the launch edge. The start point is the flop *F1* triggered by *Clk* and the end point is the port *Data*, constrained with respect to *Strobe*.

When the setup edge moved back to point *P*, the hold edge needs to be restored back to its original location. This can be done through the command:

set_multicycle_path -from [get_clocks Clk] -to [get_clocks Strobe] -hold -1

The same path can be specified as *-to [get_ports Data]*, instead of *-from [get_clocks Clk] -to [get_clocks Strobe]*. Notice the negative value of the hold multiplier.

The *Data* cannot be available at exactly the same time as the *Strobe* edge. Let us assume that the duration *AB* indicates the time during which the *Data* is expected to change. That means, till *B*, the old *Data* would be available, and the new *Data* should be available by time *A*. This has to be modeled through appropriate values on *set_output_delay*.

Let us assume the duration *P* to *A* is *1.5ns*. So, this can be specified as:

set_output_delay -clock [get_clocks Strobe] -max -1.5 [get_ports Data]

Note the negative value of the delay. This indicates that the data availability is after the reference edge.

Let us assume the duration *B* to *P* is *1ns*. So, this can be specified as:

set_output_delay -clock [get_clocks Strobe] -min 1.0 [get_ports Data]

Notice that the min value is larger than the max value. Some tools do not support this. Make sure that your tool allows this!

The same timing effect can also be specified without *set_multicycle_path*. Let us assume the clock period to be *10ns*. Without the *set_multicycle_path*, the default edge for setup check would be launch at *P* and capture at *Q*. The need for data to be available at *A* can also be looked as a requirement that the data should be available *8.5ns* before the next edge *Q*. This can be specified as:

set_output_delay -clock [get_clocks Strobe] -max 8.5 [get_ports Data]

The hold check would still be made for capture at *P*. So, min delay remains the same, namely, *set_output_delay -clock [get_clocks Strobe] -min 1.0 [get_ports Data]*.

This example showed how the same path can be specified in multiple ways. And also, the same timing effect can be achieved through different ways.

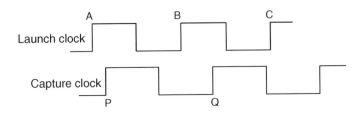

Fig. 12.9 Asynchronous clocks

12.5.3 Reset

In many ASIC designs, the master reset signal remains asserted for several cycles. So, the assertion of these signals can be declared as multi cycle paths. Assuming an active low asynchronous reset kept asserted for 3 cycles, the command would be:

set_multicycle_path -fall -from reset_n -setup 3
set_multicycle_path -fall -from reset_n -hold 2

12.5.4 Asynchronous Clocks

In Chap. 7, we have seen that asynchronous clock domain crossings are declared as *set_clock_groups -asynchronous*. The timer effectively disables such paths from any timing analysis. Thus, a path crossing clock domain can be allowed to have any amount of delay. Many designers want to put some kind of upper limit on the delay that such paths might have. This can be achieved through the commands:

set_multicycle_path -from [get_clocks C1] -to [get_clocks C2] -setup 2
set_multicycle_path -from [get_clocks C1] -to [get_clocks C2] -hold 1

When we specify a setup of 2 cycles, effectively the delay on the asynchronous path gets capped to 1 cycle (and not 2 cycles). Because the path is asynchronous, an edge pair will be considered where the launch and the capture edges are close. Thus, 1 cycle is effectively lost because of the close edges. Figure 12.9 explains this.

The default launch – capture combination is edges A and P which are very close. With a setup multi cycle of 2 (in terms of end clock), the capture edge moves to Q. The data path then gets the time from A to Q, which is almost one clock (of destination clock).

12.5.5 Large Data Path Macros

Some data paths have huge adders, multipliers, or other data path elements. Or, they might have deep levels of logic. Or, they might have a high setup requirement for the capturing device (say, a memory), or the launching device might have a high *Clk-to-output* delay (e.g., a memory). Or, the path might be on a clock which has very high frequency. In many such cases, it might be difficult for the data to meet the timing requirements of a single cycle. In such cases, the path might have to be declared as multi cycle.

12.5.6 Multimode

In Chap. 15, when we discuss multimode, we will also see one more example situation of *set_multicycle_path* command.

12.6 Conclusion

Multi cycle path provides additional relaxation to the specified paths. While specifying multi cycle paths, you should be careful to ensure:

– Unintended paths do not become multi cycle.
– The amount of additional time allowed is in line with what you had intended.

If a path is under-constrained (i.e., multi cycle specification allows a wider range for the signals to arrive) than what designed for, the device might not operate at the desired frequency.

When we move the setup edge through a multi cycle path specification, the hold edge also moves. You should check if the hold edge needs to be restored back to its original location. In most cases, it should be restored back. If you do not restore the hold edge, the design might have additional buffers in the data path, in order to increase delay to meet the increased hold requirement. This would cause wasted silicon area as well as power.

This chapter discussed multi cycle paths only in the context of timing. However, there are implications on the functional design (RTL) also, to ensure:

– Data is not lost.
– Glitches are not captured.

Chapter 13
Combinational Paths

Usually, outputs are always registered – just before being presented to the port. In many cases, inputs are also registered immediately after entering the block. In any case, most of the times, each signal entering an input gets registered at least once, before it comes out through an output port.

However, sometimes there might be paths from input to output, without encountering any register. Such paths are called combinational paths. Figure 13.1 shows an example of a combinational path.

13.1 set_max_delay

A combinational path can be constrained so that the delay on the path can be limited within an upper bound. This can be done through *set_max_delay* command. The SDC syntax for this command is:

```
set_max_delay    [-rise] [-fall]
                 [-from from_list]
                 [-to to_list]
                 [-through through_list]
                 [-rise_from rise_from_list]
                 [-rise_to rise_to_list]
                 [-rise_through rise_through_list]
                 [-fall_from fall_from_list]
                 [-fall_to fall_to_list]
                 [-fall_through fall_through_list]
                 delay_value
                 [-comment comment_string]
```

The options related to path and transition specification and comment are same as those explained in Chap. 11 for *set_false_path*, and thus a detailed explanation is omitted in this chapter. The *delay_value* specifies the upper limit of the allowed

Fig. 13.1 Combinational path

delay for this combinational path. For example, if the path shown in Fig. 13.1 is allowed to have a maximum delay of *8ns*, the corresponding command would be:

set_max_delay -from [get_ports I1] -to [get_ports O1] 8.0

13.2 set_min_delay

If the path is required to have a lower bound for the delay, the requirement can be specified through set_min_delay command. The SDC syntax for the command is:

set_min_delay   [*-rise*] [*-fall*]
                      [*-from* from_list]
                      [*-to* to_list]
                      [*-through* through_list]
                      [*-rise_from* rise_from_list]
                      [*-rise_to* rise_to_list]
                      [*-rise_through* rise_through_list]
                      [*-fall_from* fall_from_list]
                      [*-fall_to* fall_to_list]
                      [*-fall_through* fall_through_list]
                      delay_value
                      [*-comment* comment_string]

The options for *set_min_delay* and *set_max_delay* are the same in meaning and syntax. The only way *set_min_delay* differs from *set_max_delay* is that this command specifies the lower bound on the delay through the path, while *set_max_delay* specifies the upper bound on the delay. Thus, the actual delay for the path has to be somewhere between *set_min_delay* and *set_max_delay*.

Usually, in most cases, there might be no need to specify *set_min_delay*. Only in some specific situation, where some hold requirements might need a minimal delay value, there would be a need for *set_min_delay* specification.

13.3 Input/Output Delay

A combinational path can also be constrained using *set_input_delay* and *set_output_delay*. The syntax and semantics of these commands and their options are described in Chap. 9 and are not being repeated here. In this section, we will describe how *set_input_delay* and *set_output_delay* can be used to constrain the delay for a combinational path.

13.3 Input/Output Delay

Fig. 13.2 Combinational path – no interaction with clock

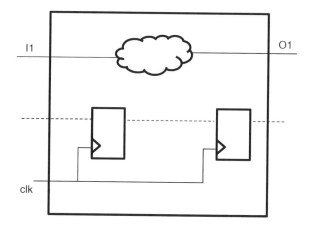

Let us say that we want to constrain the combinational path shown in Fig. 13.1 to have a maximum delay of *8ns*.

13.3.1 Constraining with Unrelated Clock

Let us also say that this same block also has a clock declaration (say *CLK*) with a period of *12ns*. This clock may not have any relationship with this combinational path, as shown in Fig. 13.2.

So, out of the period of *12ns*, a duration of *8ns* needs to be available for this combinational path. The remaining *4ns* can be distributed outside this block, through *set_input_delay* and *set_output_delay*. Say, an input delay of *3* and an output delay of *1* can be specified. The distribution of *3* and *1* among *set_input_delay* and *set_output_delay* is not important, as long as the total of input delay and the output delay specification is *4*. Thus, the following set of commands can achieve a combinational path delay of maximum *8ns*:

create_clock -name CLK -period 12 [get_ports clk]
set_input_delay -max -clock CLK [get_ports I1] 3.0
set_output_delay -max -clock CLK [get_ports O1] 1.0

The risk with this style of constraining a combinational path is, if for some reason, the clock period is modified, the combinational path delay also gets modified, even though there might be no relation between the combinational path and the clock.

13.3.2 Constraining with Virtual Clock

Instead of using a clock which is being declared for this block for some other purpose, a virtual clock can be declared, just for constraining the combinational path.

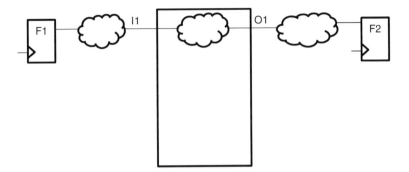

Fig. 13.3 Combinational path in the context of launching/capturing flop

We can choose whatever period we want for this virtual clock – as long as the period is more than the required max delay for the combinational path. And, the excess can be distributed among input and output delays. The following set of constraints provides one example possibility:

create_clock -name vCLK -period 15
set_input_delay -max -clock vCLK [get_ports I1] 3.0
set_output_delay -max -clock vCLK [get_ports O1] 4.0

A virtual clock has been declared with a period of *15*. Notice that there is no design object associated with the *create_clock*, thus making the clock to be virtual. With the period of *15*, there is an excess of *7ns (15–8)*. This excess has been distributed between *set_input_delay* and *set_output_delay*.

13.3.3 Constraining with Related Clock

Let us look at the same path, but this time, we also consider the circuit around this block, to show the launching and the capturing flops also – which lie outside this block. Figure 13.3 shows an example.

The launch and the capture flops lie outside the specific block. They could be lying in a different block, or they could be a part of the top-level glue logic. The input delay constraint can be specified with respect to the clock that launches data from *F1*. The input delay specified should be the delay from the launch flop till the input pin *I1*.

The output delay constraint can be specified with respect to the clock that captures data in *F2*. The output delay specified should be the delay from the output pin *O1* till the capture flop *F2*.

The clocks (which trigger *F1* and *F2*) themselves may or may not be feeding into this block. If these specific clocks are not feeding into the block, a corresponding virtual clock would need to be created.

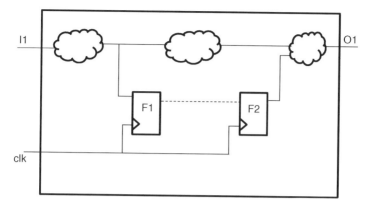

Fig. 13.4 An input/output is part of combinational as well as registered path

Among the three approaches discussed for constraining using input/output delays, this approach using the related clocks is the best. It correctly reflects the design scenario, and the allowed delay through the combinational block can then be determined automatically by the tools.

13.4 Min/Max Delay Versus Input/Output Delay

In terms of timing implication, usage of either style (viz., *set_max_delay/set_min_delay* or either of the three styles of *set_input_delay/set_output_delay*) has the same effect. However, it is preferable to constrain a combinational path using *set_input_delay/set_output_delay* combination. Let us consider the circuit in Fig. 13.4, where an input and output are part of both a combinational path as well as a registered path.

Let us say, the clock period is *15ns*. Let us say, the input arrives at *I1* at *4ns* after the clock edge. Let us say, the output *O1* has to travel for *3ns*, before it gets captured in the destination flop.

So, the delay outside the block is *7ns*. Thus, the combinational path can have a delay of *8ns* maximum. This can be specified as:

set_max_delay -from [get_ports I1] -to [get_ports O1] 8.0

Because of the input feeding into a register, there needs to be an input delay also on *I1* (in order to time *I1* to *F1* path), which will be specified as:

set_input_delay -max -clock CLK [get_ports I1] 4.0

And, in order to time *F2* to *O1* path, there needs to be an output delay also on *O1*, which will be specified as:

set_output_delay -max -clock CLK [get_ports I1] 3.0

The *set_input_delay* and the *set_output_delay* have to be anyways specified, because of *I1* and *O1*'s involvement in registered path. With these two specifications, the combinational path anyways gets constrained to *8ns*. So, there is no need for the explicit specification of *set_max_delay*.

Assuming that the *set_max_delay* is anyways specified (to *8ns*), it might be expected that these input and output delay specifications should not impact the combinational path delay (viz., *8ns*). However, in reality, the arrival time at the input and the output external delays will get counted as part of the combinational path!!!!

So, the combinational path is specified a limit of *8ns* (through *set_max_delay*). Out of that, *7ns* is contributed by the input and output delays. Thus, only *1ns* is left for the actual path. This is not what was intended. So, it is possible that for a combinational path constrained through *set_max_delay*, the effective allowed delay gets modified due to an input/output delay specification.

Now, if an input to output path is purely combinational, we would have a choice of either using only a *set_max_delay* or a combination of *set_input_delay* and *set_output_delay*. Here, since there is no need for *set_input_delay* and *set_output_delay*, it might appear as if *set_max_delay* alone is sufficient and is harmless. This is true. However, if the design gets modified so that *I1* or *O1* get involved in a registered path, they will also warrant a *set_input_delay/set_output_delay* specification. Now, this new specification of *set_input_delay/set_output_delay* ends up inadvertently modifying the max allowed delay for the combinational path.

Hence, from an ease-of-maintenance perspective, it is better to use input/output delay combination, rather than *set_max_delay*. Though the discussion was presented in terms of *set_max_delay* and *set_input_delay/set_output_delay* with *-max* specification, the same discussion holds true for *set_min_delay* and *set_input_delay/set_output_delay* with *-min* specification.

13.5 Feedthroughs

The word *feedthrough* has more than one meaning – depending upon the context. In the context of this chapter, we use the term to refer to specific types of combinational paths, wherein an input signal directly reaches an output port, without any circuit. The delay for a feedthrough path is just the wire delay inside the block. A feedthrough path often spans several consecutive blocks. The discussion mentioned in this section in the context of a feedthrough is equally applicable for other combinational paths also, if they happen to span through multiple consecutive blocks.

Let us consider a feedthrough path which spans across four consecutive blocks, as shown in Fig. 13.5.

Let us assume that the total path delay from *S*(ource) to *D*(estination) is supposed to be within *13ns*. Let us assume that the delay within each block is supposed to be maximum *2ns* and that the time of flight from one block to another block can be maximum *1ns*. The time of flight for *S* to *B1* and from *B4* to *D* also has a maximum limit of *1ns* each. So, the total path delay stays within *13ns*.

13.5 Feedthroughs

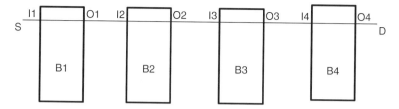

Fig. 13.5 Feedthrough path spanning multiple blocks

We will now apply constraint on all these blocks. One way is to specify the following for each of the blocks. The actual port names have to be specified in the command below:

set_max_delay -from <input port> -to <output port> 2.0

However, we have seen in the previous section that it is better to specify *set_input_delay/set_output_delay*, rather than *set_max_delay*.

For *B1*, the signal arrives at *I1* within *1ns*. After it comes out of *O1*, it has to travel for a max of *10ns* (3 blocks * 2ns each + 4 top-level routing * 1ns each).

For *B2*, the signal arrives at *I2* within *4ns* (1 block * 2ns + 2 top-level routing * 1ns each). After it comes out of *O2*, it has to travel for a max of *7ns* (2 blocks * 2ns each + 3 top-level routing * 1ns each).

For *B3*, the signal arrives at *I3* within *7ns* (2 blocks * 2ns each + 3 top-level routing * 1ns each). After it comes out of *O3*, it has to travel for a max of *4ns* (1 block * 2ns + 2 top-level routing * 1ns each).

For *B4*, the signal arrives at *I4* within *10ns* (3 blocks * 2ns each + 4 top-level routing * 1ns each). After it comes out of *O4*, it has to travel for a maximum of *1ns* (1 top-level routing * 1ns).

Assuming a clock *CLK* has already been created with a period of *13ns*, the constraints would be specified as:

For *B1*:
set_input_delay -max -clock CLK [get_ports I1] 1.0
set_output_delay -max -clock CLK [get_ports O1] 10.0

For *B2*:
set_input_delay -max -clock CLK [get_ports I2] 4.0
set_output_delay -max -clock CLK [get_ports O2] 7.0

For *B3*:
set_input_delay -max -clock CLK [get_ports I3] 7.0
set_output_delay -max -clock CLK [get_ports O3] 4.0

For *B4*:
set_input_delay -max -clock CLK [get_ports I4] 10.0
set_output_delay -max -clock CLK [get_ports O4] 1.0

It should be seen that the delay through each of the block gets constrained to a max of *2ns*.

13.5.1 Feedthroughs Constrained Imperfectly

Let us say that after the above is tried, for some reason, the timing for *B2* could not be met. Say, the delay for this block could not be reduced below *2.5ns*. So, for some other block, the delay has to be reduced. Say, for *B4*, the delay can be reduced to *1.5ns*. So, the total delay for the whole feedthrough path remains the same. However, the *set_input_delay* and *set_output_delay* for the individual blocks would need to be modified. This will change the arrival time for *B3* and *B4* (arrives *0.5ns* later), and external time on the output side of *B3* and *B2* (external required time is *0.5ns* lesser), thus changing many input/output delays – including for blocks like *B3* – for which there was no change in the routing/delays inside it.

Often, for such paths, where a feedthrough passes through several blocks, many designers do not necessarily specify the actual arrival time for input and external required time for output. Rather, they would choose a pair of input and output delay values such that the delay inside the block is the desired value. For the example described, the output delay for *B2* would be reduced by *0.5ns*. And, the input delay for *B4* would be increased by *0.5ns*. The constraints for *B3* would be left unchanged. Though, it still means *2ns* inside *B3*, the *set_input_delay/set_output_delay* no longer represents the actual time of arrival or the actual time needed to travel after coming out of the block.

Usually, on some very high-performance designs, e.g., processors – several blocks are designed as hard-IPs. In order to not impact the timing due to routing on higher layers, these IPs provide feedthrough paths. A path going from one block to another could feedthrough several IPs. Such designs might have such characteristics, where the input and output delays are different from the actual values.

13.6 Point-to-Point Exception

As shown in Sect. 13.4, usually for port to port paths, *set_input_delay* and *set_output_delay* combination is preferred compared to *set_max_delay*, even if the path is purely combinational.

Sometimes, a path segment on an entire path inside the design might need to be constrained to a special value. *set_max_delay* might be very useful for such point-to-point exceptions. Figure 13.6 shows a simple double-flop synchronizer due to an asynchronous clock domain crossing.

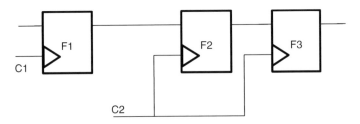

Fig. 13.6 Simple double-flop synchronization

The path from *F1* to *F2* should not be timed. This can be achieved through either *set_clock_groups* or *set_false_path*. Or, a user might put a *set_multicycle_path*. The path from *F2* to *F3* gets constrained due to clock period defined for *C2*. Usually, no logic is put between *F2* and *F3*. It is possible that placement and routing tools can place these flops far apart or take a long route, since they see the complete clock period available for this path. If the path from *F2* to *F3* is long, then the whole purpose of putting a direct path without any other logic is lost. The effectiveness of the synchronization is reduced (means MTBF (mean time between failures) will not increase as much as desired). Designers usually want that the delay from *F2* to *F3* should be very small – much smaller than the allowed clock period. So, they will constrain this path using *set_max_delay*, e.g.,

set_max_delay -from F2 -to F3 <value>

13.7 Path Breaking

Before applying a *set_max_delay* or a *set_min_delay*, the designer should understand that if the constrained portion does not start from a timing start point or end at a timing end point, these constraints break the path, at both ends of the path segment. Figure 13.7 shows a design excerpt, which has several paths, namely, $F1 \rightarrow F3$; $F1 \rightarrow F4$; $F2 \rightarrow F3$; $F2 \rightarrow F4$.

Each of these paths will be timed. However, let us say, for some reason, a *set_max_delay* or a *set_min_delay* is specified *-from I1/Z -to I2/A*. Now, a timing path that starts from *F1* (or *F2*) will stop at *I1/Z*, which has become a new start point. It is very important to understand this path breaking nature of *set_min_delay* and *set_max_delay*. Because of this characteristic, the paths $F1 \rightarrow F3$; $F1 \rightarrow F4$; $F2 \rightarrow F3$ and $F2 \rightarrow F4$ will not be timed any more.

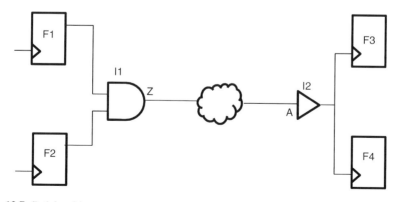

Fig. 13.7 Path breaking

13.8 Conclusion

A combinational path can be constrained using *set_min_delay* and *set_max_delay*. If the paths span from an input port to an output port, it is better to constrain the path using *set_input_delay* and *set_output_delay* combination. In general, since in most cases, the interest is in making sure that the delays are lesser than a desired value, so, *set_max_delay* is used more often than *set_min_delay*.

Chapter 14
Modal Analysis

Today's designs are very complex. They are "System on a Chip" in the real sense. The same chip performs multiple functions at different points of time. Within the chip also, there are portions in the design which behave one way in one use mode and behave differently in another use mode.

14.1 Usage Modes

A portion of the design might have one requirement for one kind of operation. And, for a different kind of operation the same portion of the design might have a different requirement.

The best example could be a design in the video entertainment segment. In the video world, user experience is a major requirement. In order to provide a real-like user experience, performance becomes the key factor. On the other hand, when the user is not using the device for a video application, performance is no longer important. Rather, it is more important to conserve battery life (thus, power) – even if performance has to be scaled down significantly.

Thus, parts of the device could have changing requirements – depending upon what mode the device is currently in. Individually, each part of the design has to meet the requirements of each of the individual modes.

14.2 Multiple Modes

For the sake of simplicity, let us say, a device has two major usage scenario – represented as modes *M1* and *M2*. Let us further assume, there are two parts – *P1* and *P2* in the design, for which the timing requirements change depending upon whether the device is being used in mode *M1* or *M2*.

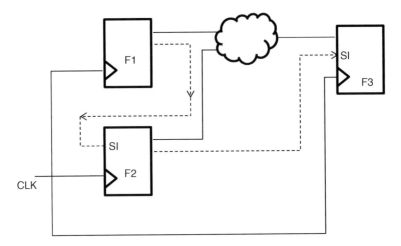

Fig. 14.1 Functional and test mode

Now, *P1* has to meet the requirements of both the modes, *M1* and *M2*. Thus, *P1* has to be designed to meet the most restrictive of the requirements. Similarly, *P2* has to be designed to meet the most restrictive of the requirements among *M1* and *M2*. However, there is no situation when *P1* will be operating in mode *M1* while *P2* would be operating in mode *M2*. Both *P1* and *P2* will together operate in mode *M1* or in *M2*.

Let us consider the circuit shown in Fig. 14.1.

The paths shown with solid lines indicate functional paths – which are active when the circuit is in normal operation. The paths shown with dotted lines indicate scan paths – which are active during Scan Shift. The same *CLK* port is used for *SystemClock* during functional mode and *TestClock* for Scan mode.

The *SystemClock* usually operates at a higher frequency, say a period of *10ns*; while *TestClock* usually operates at a lower frequency, say a period of *40ns*.

The path *F1 → F3* is a functional path and should meet the timing corresponding to *10ns* period. The path *F1 → F2/SI* is a scan path and should meet the timing corresponding to *40ns*.

We need to specify *SystemClock* so that path *F1 → F3* gets timed correctly. We also need to specify *TestClock* so that path *F1 → F2* gets timed. Since both *SystemClock* and *TestClock* share the same port, both clocks will be declared at the same location – which is *CLK* port. Now, during timing analysis, each of the paths will get analyzed corresponding to both *SystemClock* as well as *TestClock*. Thus, path *F1 → F2* will be timed corresponding to *SystemClock* also – which is an overkill. The path will be forced to meet *10ns*, when *40* is good enough.

However, at any time the device will be in only one mode – either it will be in normal operation or it will be under scan mode. If it is in functional mode, the path *F1 → F2* is not of interest. And, when the path *F1 → F2* is of interest, the device is in scan mode.

In such situations, we can define two different modes for the device. We could define a functional mode. In this mode, it analyzes paths $F1 \rightarrow F3$ and $F2 \rightarrow F3$ using *SystemClock*. And, we could define another mode for scan. In this mode, it analyzes path $F1 \rightarrow F2$ using *TestClock*.

14.3 Single Mode Versus Merged Mode

A user could write the constraints for each mode individually, or write a set of constraints which are combined for multiple modes.

Usually front end designers who write the RTL code to represent the functionality find it easier to comprehend the design in terms of various functional modes. It comes more naturally for them to think of the design in terms of functional mode. Hence, they prefer to write the constraints for each mode individually.

In Sect. 14.6, we will see some of the challenges that arise due to individual mode constraints. Because of those challenges, the backend designers tend to merge the constraints. For them, the design is usually less about the functionality. They look at the design as a network of logic elements, and don't tend to think in terms of individual functional modes.

14.4 Setting Mode

When an SDC represents a single mode, certain points in the design can be fixed at specific values that are unique characteristics of that mode. The SDC command for setting a specific value is *set_case_analysis*. The SDC syntax for the command is:

set_case_analysis value port_pin_list

where, value can be *0/1/rising/falling*.

The command fundamentally conveys that for the current analysis assume that a given object is at the specified value or transition.

For putting a device into a specific mode, sometimes just one *set_case_analysis* might be sufficient. And, sometimes, a set of several *set_case_analysis* might be needed to put the device into a specific mode.

Figure 14.2 shows the same circuit as Fig. 14.1 – but with some more details.

For this example to be analyzed in the functional mode, the clocks will be declared as:

create_clock -name SysClk -period 10 [get_ports CLK]

In addition, we should apply

set_case_analysis 0 [get_ports SE]

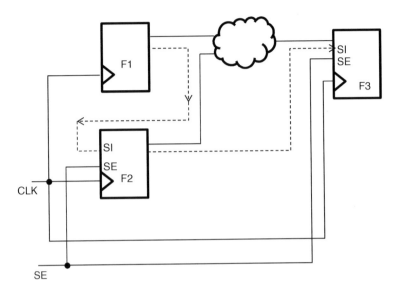

Fig. 14.2 Scan pins shown for the previous circuit

The flop models contain the information that when *SE* pin is *0*, only *D* pin can be sampled. Thus, the paths to *SI* pin will not be analyzed, but the paths to *D* pin will be analyzed. Thus, path $F1 \rightarrow F2$ will not be analyzed in this mode, because this path reaches *SI* pin of *F2*.

On the other hand, if we want the example to be analyzed in the scan mode, the corresponding commands would be:

create_clock -name TstClk -period 40 [get_ports CLK]
set_case_analysis 1 [get_ports SE]

Again, because flop model contains the information that when *SE* pin is *1*, only *SI* pin can be sampled. Thus, the paths to *D* pin will not be analyzed but the paths to *SI* pin will be analyzed. Thus path $F1 \rightarrow F2$ will be analyzed in this mode. Also, the path from $F2 \rightarrow F3$'s *SI* pin will also be analyzed.

Let us consider a block, which has several possible modes of operation. There is a configuration register of 8 bits whose setting decides the specific mode of operation. In such a case, all 8 bits of the register might need to be set in order to put the device in the mode of interest. Example commands could be something like:

set_case_anlaysis 0 [get_pins config_reg[0]/Q]
set_case_anlaysis 1 [get_pins config_reg[1]/Q]
set_case_anlaysis 1 [get_pins config_reg[2]/Q]
set_case_anlaysis 0 [get_pins config_reg[3]/Q]
set_case_anlaysis 1 [get_pins config_reg[4]/Q]
set_case_anlaysis 0 [get_pins config_reg[5]/Q]
set_case_anlaysis 1 [get_pins config_reg[6]/Q]
set_case_anlaysis 0 [get_pins config_reg[7]/Q]

14.4 Setting Mode

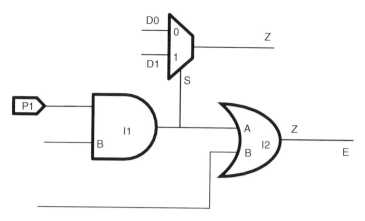

Fig. 14.3 Case analysis impact on paths being timed

In order to decide the *set_case_analysis* settings, we first need to decide the mode of operation for which the analysis has to be done. After that we need to decide the control pins/ports which influence the mode of operation for the device. Then, we need to specify those pins/ports to be at the values which will put the device into that mode of operation. Usually, *set_case_analysis* is applied only on ports or on register output pins. Usually, the register pins are used to set values on configuration registers. Even though, the syntax allows the values to be applied on any pin.

set_case_analysis command prevents certain paths from participating in timing analysis. This prevention happens in multiple ways. First, the specific pin being constant does not originate or transmit any transition. Second, the values specified through *set_case_analysis* propagate to the rest of the design – thus putting additional constants in the design. Third, these constants (either specified directly through *set_case_analysis* or after propagation) disable certain paths from being timed. Circuit shown in Fig. 14.3 provides an example of how the values applied through *set_case_anlaysis* propagate and disable certain paths from participating in timing analysis.

Let us assume that in order to set the device into a specific mode, the following constraint has been specified:

set_case_analysis 0 [get_ports P1]

So, any path involving a transition on *I1/A* no longer participates in timing analysis, as *I1/A* is always held at a constant value. A transition on *I1/B* will also not reach *I1*'s output pin. Hence, any path involving a transition on *I1/B* will also not participate in timing analysis. The value of *0* on *I1/A* propagates to the output of the *AND* gate and then to the *Sel* pin of the *MUX*. Because *MUX*'s *Sel* pin is held at constant, so, paths through this pin will also not be timed. And, paths through *D1* pin of the *MUX* will also not participate – because, *MUX*'s *Sel* pin being at *0* means *D1* will not reach the output. Only the paths through *D0* to output of *MUX* and the *I2/B* to *I2/Z* will be timed.

14.5 Other Constraints

Using the *set_case_analysis* command we can set specific points in a design to fixed logical values which characterize a specific mode of operation. Besides setting the logical values, the constraints for a specific mode also means setting other constraints like clock definitions, input and output delays etc. also which are specific to the intended mode of operation. Thus for the same input port, in one mode, it could have one input delay and in another mode, it could have another input delay.

For example, let us say, an input port receives data with respect to one clock in one mode of operation and data with respect to another clock in another mode of operation. In this case, for each mode, the input would be constrained with respect to one clock only (corresponding to that mode).

Or, an input might receive signals at different time in different mode of operation. In such cases also, the input delay specified in a specific mode is usually the value corresponding to that mode of operation.

In short, while writing constraints for a specific mode, the constraints are written as if that is the only mode in which the device will operate.

14.6 Mode Analysis Challenges

The advantage of analyzing individual modes is that certain timing paths which are never possible in the actual device operation get excluded from timing analysis, e.g., referring to Fig. 14.1, $F1 \rightarrow F2$ path in functional mode need not be timed. However, mode analysis also has its own challenges.

14.6.1 Timing Closure Iterations

Let us consider a design with four different modes – *M1*, M2, M3, and *M4*. The design has to meet the timing for each mode individually. The designer will synthesize the design for any one mode – say *M1*. Now, if timing analysis is done for *M1*, the timing might be clean. The same design also needs to be analyzed and made timing clean for mode *M2*. It is possible that certain paths which are applicable in *M2* were not to be analyzed in *M1*. These paths might potentially fail timing, when subjected to timing analysis in mode *M2*. So, some fixes will have to be made into the design – so that these paths also start meeting the timing. After mode *M2* is also timing clean, the timing analysis will need to be done for mode *M3*. Once again, it is possible that some paths valid in mode *M3* might fail the timing. So, some fixes will have to be made once again – so that these paths also start meeting the timing. Similarly, analysis will be needed for mode *M4*, which might cause some more fixes.

By now, each of the modes has been individually analyzed and where needed, fixes were also made. However, as part of making these fixes, the design has been altered. Any timing analysis done before the design was last altered is no longer valid. Thus, timing analysis will need to be done once again for each of the modes after the last update.

When we redo the timing analysis for mode $M1$ – it is possible that some path might fail timing now. During the previous iteration of mode $M1$, this same path was meeting the timing. As part of making fixes for other modes, it is possible that this path was diverted through a longer route. And, as part of fixing this, it is possible that some other path is diverted through a longer route – which could potentially cause some other mode to fail.

Thus, the whole process goes into a loop, where a fix of a path in one mode causes another path applicable in another mode to have broken timing. The fundamental problem is that implementation tools have been made to see only a subset of the paths at any given time and they try to meet only those paths. In the process, they might deteriorate paths which are not being seen by them in the current mode. At this time, the tools are not able to see that the paths which are being deteriorated could be important in another mode, causing a failure in that other mode.

Today, with complete systems on a single chip, many of the designs have more than ten modes. So, there are too many analysis required; and there is always a risk of this going into a loop. This loop is commonly referred as timing closure iterations.

As a solution to this problem, many designers try to combine various modes into a single hypothetical mode. The concept is called Mode Merge. We will read more about it in the next chapter.

14.6.2 Missed Timing Paths

We have seen earlier in this chapter, that when we apply some *set_case_analysis*, certain paths get excluded from the timing analysis. However, it is expected that each timing path in the design is there for some specific purpose, and each path should be required to meet some timing in some mode or other. Typically, a design has millions of timing paths. In each mode, thousands of paths may get excluded from timing analysis.

However, each timing path should get covered in at least some mode or the other. There is no good way of knowing that each path been covered. Effectively, there is a risk that some specific path got excluded from each of the mode settings and was not timed at all in any of the modes. This could happen because the *set_case_analysis* used for some mode were incorrect; or because a certain mode was not considered for analysis.

14.7 Conflicting Modes

Because of the problem related to timing closure iteration mentioned above, sometimes, users will set different parts of the design in different modes, which might even be conflicting. They do this so that not many different modes have to be created. Let us consider the circuit shown in Fig. 14.4.

Fig. 14.4 Conflicting mode settings

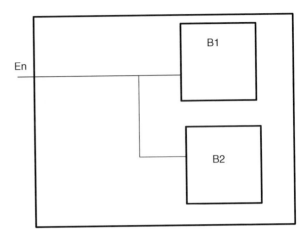

Let us assume that the design pin *En* controls the operational mode of the design. Hence, for two different runs, the *case_analysis* settings would be:

set_case_analysis 0 [get_ports En]

and

set_case_anlaysis 1 [get_ports En]

As mentioned in the previous section, this will mean doing timing analysis twice. Sometimes, this could also cause iterations through the implementation tools.

Let us further assume that the block *B1* has the most restrictive timing when its pin *En* is at *0*, while the block *B2* has the most restrictive timing when its input pin *En* is *1*. In such a case, some designers prefer to specify:

set_case_analysis 0 [get_ports En]
set_case_anlaysis 1 [get_pins B2/En]

It should be noticed that it is never possible to have the above situation in the design, since *B2/En* is being driven directly by the *En* port. However, this allows the design to be put into the most restrictive situation and do the analysis only once.

Similarly, sometimes, conflicting values are set at flops. A *set_case_analysis* set somewhere could propagate a *0 at* a specific flop's input, while the flop's output might have a *1* set at it. Again, not something that is actually possible in the design. However, this covers the situation, where the timing was supposed to be most restrictive in one condition till the flop; and after the flop it is most restrictive in an opposite condition.

Merging of several SDC files belonging to different modes into one SDC file is dealt in more detail in the next chapter. This example has been given mostly to show that sometimes, mode settings could be made in a conflicting manner, even though, logically these situations may never occur in the actual design.

14.8 Mode Names

It should be noted that several times in this chapter, we refer to mode names. However, SDC does not provide any mode naming convention/command. Individual tools might still provide a mechanism to provide a name to the mode. In the context of SDC, any name for the mode is mostly for understanding of the user – as to which functionality does he want to cover, by the corresponding *set_case_analysis* commands.

14.9 Conclusion

Mode analysis allows a user to restrict the analysis to specific operational situations only, rather than considering all possible combinations of paths and situations, some of which might never happen in the design. Mode analysis makes it easier for the designer to write constraints only for specific operational modes. However, dealing with only a subset of paths for one mode, without any consideration for the other paths, which will be meaningful in other modes often causes a long iterative loop through the timing closure.

For most designs, front end designers generate the SDC file specific to individual modes. However, the backend engineers merge several modes into one constraint file, so that the implementation tools can look at the whole set of requirements in one go.

Chapter 15
Managing Your Constraints

As the complexity of designs increases there is a need to accurately model timing constraints for early design closure. When done correctly, they help not only to achieve faster timing closure but also reduce iterations between front-end and back-end teams. In reality constraints are constantly being tweaked as the design is being pushed from RTL to post layout. This requires the design to be partitioned appropriately and the constraints to be managed effectively so that the design intent is preserved at every step.

During the design development chip architect makes a call on how the design needs to be partitioned, optimized, and assembled. Depending on the complexity of the design and the level of integration involved designers may choose one of the following three flows

1. Hierarchical top-down methodology
2. Bottom-up methodology
3. Bottom-up top-down (hybrid) methodology

Further depending on the number of modes per block or chip, the designer may want to reduce his timing closure iterations by merging modes to manage his constraints effectively.

15.1 Top-Down Methodology

In this methodology the entire chip is considered one single design unit and constraints are applied at the top level and the design is synthesized as a single unit. The advantage of this methodology is that it boasts of a simple use model and makes optimization relatively simple and worry free step. However this model is not very scalable. For really large designs this is dependent on the implementation tool capacity and the hardware on which it is running. Further any minor change in any part of the design results in a complete re-implementation.

15.2 Bottom-Up Methodology

This is a variant of the hierarchical top-down flow, in the sense that it uses divide and conquer approach to partition a chip into sub-chips and sub-chip into blocks. Constraints are created for each sub-chip or block, which are then analyzed similar to the top-down flow. Each block is optimized separately based on its own constraints and integrated at the sub-chip or chip level. This has the advantage that it makes the design methodology scalable for large designs. An incremental change to a block doesn't require the entire design to be synthesized. However this methodology may result in integration issues at the top level. A block which when optimized separately may meet its timing, but when integrated in the context of the sub-chip or chip may fail timing. Similarly critical paths at intermediate level or top level of hierarchy may not be apparent as critical at the block level. This is because the constraints created at the block level have no visibility to the constraints at the top level and the constraints of the adjoining blocks it is going to interact with. This increases the iteration when multiple blocks are integrated and the design tries to meet inter block and intra block timing requirements.

Further at the sub-chip level, to create constraints, designers may resort to propagation of block level constraints. This is not straightforward, as there could be conflicting clock and case analysis constraints from multiple blocks to deal with. Figure 15.1 shows an example of such a conflict.

In this figure, the block level constraints are such that port *P1* is constrained in block *B1* to have a value of *1*, while *P1* of block *B2* is set to *0*. Then at the chip level there is feed-through path from *B1/P1* → *B1/O1* → *B2/P1*, which causes these constraints to conflict. Why is this an issue? *B1* has been timed for a mode when its port

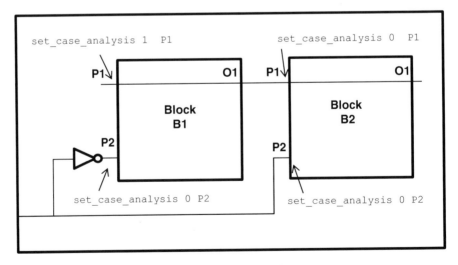

Fig. 15.1 Bottom-up constraints propagation causing conflict

15.2 Bottom-Up Methodology

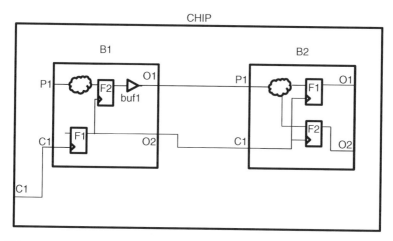

Fig. 15.2 Bottom-up constraints propagation

P1 is at *1*. Similarly *B2* has been timed for a mode when port *P1* is at *0*. But in the actual functioning of the chip, since they cannot have values *1* and *0* simultaneously, the mode is hypothetical at chip level. One of the blocks may exhibit worse timing, compared to what it showed when timed individually. Similarly, if port *P2* in block *B1* and *B2* are constrained to value *0*, then from the logic it is clear that both ports cannot have the same value, since there is an inverter in the path to *P2* of block *B1*.

In the examples above, we have kept the port names the same across blocks just to make it easy to understand. Many times the port names may not be the same, while there is a relation among the ports. Such cases are quite difficult to detect, through text-based human review of constraints.

When Constraints are propagated upwards in a bottom-up flow, there are three kinds of transformations that are possible.

1. Constraints at block level are applied as is at chip level with only the hierarchy updated.
2. Constraints at block level are modified to reflect the SoC context.
3. Constraints are dropped, since they don't make sense in the SoC context.

Let us consider the example in Fig. 15.2. In this example the block *B1* defines a clock for *C1* and generated clock *GC1* in its own constraints file. Block *B2* defines a clock *C1* in its own constraints file.

Constraints of block B1
create_clock -name C1 -period 10 [get_ports C1]
create_generated_clock -name GC1 -divide_by 2 -source C1 [get_pins F1/Q]
set_false_path -from F2

Constraints of block B2
create_clock -name C1 -period 20 [get_ports C1]
set_false_path -from P1 -to F1

When *B1* and *B2* are integrated at chip level, the clock *C1* and *GC1* (with hierarchy updated) are retained. Since *GC1* effectively drives the clock modeled by *C1* of block *B2*, an additional constraint is not required for *B2*'s clock. Further the *set_false_path* in block *B1* should be simply transformed to reflect the new hierarchy.

Constraints of CHIP based on propagation
create_clock -name C1 -period 10 [get_ports C1]
create_generated_clock -name GC1 -divide_by 2 -source C1 [get_pins B1/F1/Q]
set_false_path -from B1/F2

At chip level we don't need the false path from *P1* to *F1* in block *B2*, since the false path from *B1/F2* covers that path as well. Therefore at chip level this can be dropped. It is important to note sometimes a mere hierarchy manipulation may not be enough, if at the SoC level additional paths become false. It is always advisable that such anomalies, if any, should be resolved based on knowledge of the design.

If the designer decides to create the chip level constraints manually, he would need to validate consistency between chip and block constraints. This step would ensure that for each constraint at the top level there is no conflict with the constraints at the block level.

- Let us consider Fig. 15.3 which shows an example of a block level constraints and the corresponding chip level constraint. In this case, the clock constraints at the block and chip level are not consistent. The block is constrained with a *10ns*

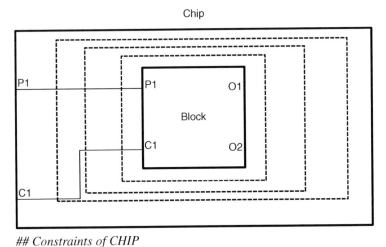

Constraints of CHIP
create_clock -name C1 -period 8 [get_ports C1]
set_input_delay 6 -clock C1 [get_ports P1]

Constraints of block
create_clock -name C1 -period 10 [get_ports C1]
set_input_delay 4 -clock C1 [get_ports P1]

Fig. 15.3 Chip vs. block constraints validation

clock, however at the chip level it is being driven by an *8ns* clock. The block may not have been designed to run with a faster clock and could result in the chip not to function properly. Similarly the input delay at chip level is more than input delay at the block level. This indicates inconsistency in input delays as the delay at the block port cannot be lower than that at the chip boundary, due to finite time of flight delay between chip level port *P1* and its connection at the block level *P1*.

Hence in the bottom-up methodology, designers may need to tweak both the subchip and block level constraints to reach a convergence on meeting both the chip and block level timing requirements. As explained before, this is because the block level constraints are supposed to be created keeping in view its surroundings in the context of the chip where it will be integrated. However there is a possibility of an error, because constraints keep changing due to continuous give and take between various blocks. To circumvent this problem, designers these days use a hybrid of the bottom-up and top-down flow, also referred to as the Bottom-up Top-down methodology.

15.3 Bottom-Up Top-Down (Hybrid) Methodology

This methodology includes a context sensitive optimization of the block where constraint for the lower level blocks is derived from the chip level constraints. The block is then timed using the traditional bottom-up flow. This methodology ensures that the block constraints consider the impact of the adjoining blocks which will be used during the chip integration. The interface level constraints for the block include information about the block, the arrival time at the input, the required time at the output and other parameters like capacitance, load, driving cell, etc. for accurate context sensitive optimization. This methodology also ensures that delays are correctly split across blocks and in a balanced fashion. This process is called budgeting.

The timing budgeting step ensures the following:

1. Delays are allocated between blocks in order to meet top level timing requirements. This allocation could be based on either a fixed proportion of clock period, or a ratio depending upon the number of logic levels.
2. After the initial budgets are allocated to the blocks and blocks are synthesized, some blocks may have positive slack and some may have negative slack. The block constraints are refined and slack is redistributed on the input and output delay of the blocks in such a way that positive slack paths don't become negative and negative slack paths don't become any worse.
3. The process can be iterative, but incremental changes require only the effected blocks to be resynthesized and not the entire design.

If a block has more than one instance, then each instance may have to be synthesized with different set of constraints to model the context sensitive nature of optimization. This process where each instance is treated differently is called uniquification.

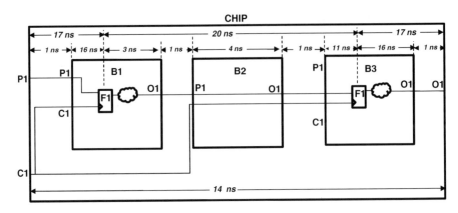

Fig. 15.4 Budgeting chip constraints

Figure 15.4 shows an example of how constraints at the top level are budgeted to the blocks.

In this example, the clock at the chip level has a period of *20ns*. Let the input delay be *3ns* and output delay be *3ns* as well. The chip level constraints are shown below.

Constraints of chip
create_clock -name C1 -period 20 [get_ports C1]
set_input_delay 3 -clock C1 [get_ports P1]
set_output_delay 3 -clock C1 [get_ports O1]

This means there is *17ns* (clock period – chip input delay) time available from the port *P1 at* chip level and Flop *F1* in block *B1*. Further the time from flop *F1* in block *B1* to flop *F1* in block *B3* is *20ns*. Let us further assume that all the interconnect delays are *1ns*. Also let the delay inside the feed-through block be *4ns*. This means that the time available inside block *B3* from its port to the flop *F1* is *11ns*. Finally the path between flop *F1* inside block *B3* and output *O1 at* chip level has a *17ns* (clock period – chip output delay) time available. This will result in the block constraints as shown.

Constraints of block B1 Post Budgeting
create_clock -name C1 -period 20 [get_ports C1]
set_input_delay 4 -clock C1 [get_ports P1]
set_output_delay 17 -clock C1 [get_ports O1]

Constraints of block B2 Post Budgeting
create_clock -name C1 -period 20 ## Virtual clock
set_input_delay 4 -clock C1 [get_ports P1]
set_output_delay 12 -clock C1 [get_ports O1]

Constraints of block B3 Post Budgeting
create_clock -name C1 -period 20 [get_ports C1]
set_input_delay 9 -clock C1 [get_ports P1]
set_output_delay 4 -clock C1 [get_ports O1]

Now if allocation of these budgets causes any block to have negative slack, then delay from the block with positive slack can be redistributed. For example, if block *B1* has *1ns* negative slack and *B2* has a *1ns* positive slack, the delay can be redistributed. This will result in the constraints for *B1* and *B2* to change. The user would need to re-synthesize only *B1* and *B2*. Since *B3* is unaffected, no more change needs to happen.

Constraints of block B1 Post Budgeting
create_clock -name C1 -period 20 [get_ports C1]
set_input_delay 4 -clock C1 [get_ports P1]

Constraints that changed
set_output_delay 16 -clock C1 [get_ports O1]

Constraints of block B2 Post Budgeting
create_clock -name C1 -period 20 ## Virtual clock
set_input_delay 5 -clock C1 [get_ports P1] ## Constraints that changed
set_output_delay 12 -clock C1 [get_ports O1]

Constraints of block B3 Post Budgeting – Remains unchanged
create_clock -name C1 -period 20 [get_ports C1]
set_input_delay 9 -clock C1 [get_ports P1]
set_output_delay 4 -clock C1 [get_ports O1]

The hybrid approach is advantageous because it enables faster convergence at the block level with respect to chip level, thereby reducing iterations during the integration phase.

All this analysis is for a single mode of operation. In reality a chip has many modes of operation and user would end up doing this kind of analysis for each mode, which could be tedious. Let us now try to understand how you can manage constraints, when you are dealing with multiple modes.

15.4 Multimode Merge

As described in Chap. 14 a chip can have multiple operational modes like functional mode with fast clock (say for high-end graphics application), functional mode with slow clock (for normal operation), a test mode, a sleep mode for power savings and in some cases a debug mode to run diagnostics. STA is typically performed for different PVT (Process, Voltage, and Temperature) corners also. These could be Worst Case (Slow process, Low Voltage, and High Temperature), Best Case (Fast process, High Voltage, Low Temperature), Typical Case (Typical process, Nominal Voltage, and Nominal Temperature). Just between the operational modes and the PVT corners, a single chip might need to be analyzed for 18 (6 modes×3 corners) different modes. When you do signal integrity analysis, you would need to consider the impact of different parasitic interconnect corners, which could add the third dimension to the number of modes. Thus trying to run a design on all these modes to meet timing and on each one of them is a giant task by itself and to reach convergence in a way that all modes meet timing is extremely difficult and painfully iterative, as explained in Sect. 14.6.1. This kind of analysis is generally referred to as Multimode-multicorner (abbreviated as MMMC) analysis.

We saw in the previous chapter that one of the challenges of MMMC analysis was that incrementally modifying constraints of a mode may cause unwanted timing failure on a mode on which timing may have been closed in the previous iteration. STA and implementation tools today provide MMMC support, wherein they are able to simultaneously consume the constraints for all modes and then perform a unified optimization. However this can be very runtime intensive and designers may not get a meaningful result in a reasonable time. Besides, most tools do not dump out the "envelope" constraints – the effective constraints that they were considering, which cover all the individual modes/corners. So, designers are not able to review whether the analysis has really considered all corners and modes. Hence often, designers would constrain their blocks by merging all the modes to create a super-mode constraint file.

Mode merge is the technique in which constraints of different operational modes are combined with the sole aim of consolidating into a single mode. This single constraint is a hypothetical mode that models the union of all modes. Typically such a mode would facilitate faster timing closure as it models the constraints in a rather pessimistic fashion. Instead of trying to painfully meet timing for each mode individually, the design can meet timing in the merged mode thereby saving implementation cycles.

Typically in a hierarchical flow, the designer tries to create merged mode constraint for the block design. However the chip level constraints are created for each mode and MMMC analysis is done at signoff. Though there is nothing that prevents any deviation from this, it is generally accepted practice today in the industry based on what the tools support. Of course, this can change in future.

Most designers do merge of multiple modes based on the understanding of the constraints. They typically adhere to the following guidelines while merging:

- Merging must not under-constrain any path or object any more than in any of the individual modes. Effectively merged constraints should not be optimistic than any of the individual modes.
- Merged constraints must be easily understood by designers.
- Merged constraints should not result in too many commands or expand to too many paths as it can overwhelm the implementation tools. For example, a brute force way of merging would be to enumerate every path and constraining it in a way that it would have worked. But this will result in over hundred thousand and in some cases over million paths, which can be beyond the tool's ability to handle. Further this volume of constraints may not be maintainable or reusable by designers.
- Theoretically you can merge all modes into a single mode, but that is generally not done. Merging is limited to a smaller set so as to keep a balance between excessive pessimism, readability, not needing too many exceptions etc.

15.4 Multimode Merge

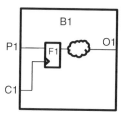

Fig. 15.5 Picking pessimistic clock

The following examples illustrate a few scenarios on how to merge multiple modes. Once the concept is understood, it should be possible for a designer to apply these/similar principles and merge the constraints for various modes.

15.4.1 Picking Pessimistic Clock

Let us consider the example in Fig. 15.5. In this example, in mode 1, the clock has a period of *10ns*. In mode 2, the clock has a period of *5ns*. In the merged mode, we could pick the clock with *5ns* period so as to not under-constrain any path more than in the individual modes.

Mode1
create_clock -name C1 -period 10 [get_ports C1]

Mode2
create_clock -name C1 -period 5 [get_ports C1]

Merged_Mode
create_clock -name C1 -period 5 [get_ports C1]

15.4.2 Mutually Exclusive Clocks

Let us consider the example in Fig. 15.6 which shows a clock mux structure with conflicting constraints in two different modes.

Mode1
create_clock -name C1 -period 10 [get_ports C1]
set_case_analysis 0 mux1/S

Mode2
create_clock -name C2 -period 40 [get_ports C2]
set_case_analysis 1 mux1/S

In this example, in mode 1, the clock *C1* has a period of *10ns*, and no clock is defined for *C2* and select line of the mux is set to *0*. In mode 2, no clock is set on

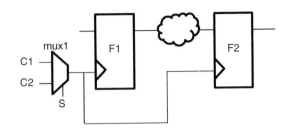

Fig. 15.6 Mutually exclusive clocks

C1, the clock *C2* has a period of *40ns* and the select line of the mux is set to *1*. While merging, since the *set_case_analysis* on the select lines are conflicting, it needs to be dropped. Both clocks *C1* and *C2* will be retained in the merged mode, since flops *F1* and *F2* need to be timed with both clocks to model both modes. However any interaction between *C1* and *C2* (anywhere else in the design) will have to be nullified. This can be achieved by defining a *set_clock_group -logically_exclusive* between *C1* and *C2*.

Merged_Mode
create_clock -name C1 -period 10 [get_ports C1]
create_clock -name C2 -period 40 [get_ports C2]
*set_clock_group -logically_exclusive -group C1 *
-group C2

While specifying the logically exclusive relationship, it should be ensured that these clocks are not interacting anywhere other than the fanout cone of the mux. The following section shows how to apply the constraints correctly, if these clocks also interact somewhere other than the fanout cone of the mux.

15.4.3 Partially Exclusive Clocks

Let us consider the example in Fig. 15.7.

Mode 1
create_clock -name C1 -period 10 [get_ports C1]
create_clock -name C2 -period 20 [get_ports C2]
set_case_analysis 0 mux1/S

Mode 2
create_clock -name C1 -period 10 [get_ports C1]
create_clock -name C2 -period 20 [get_ports C2]
set_case_analysis 1 mux1/S

In this example, in mode 1, the clock *C1* has a period of *10ns* and clock *C2* has a period of *20ns*. The select line of the mux is set to *0*. However clocks *C1* and *C2*

15.4 Multimode Merge

Fig. 15.7 Partially exclusive clocks

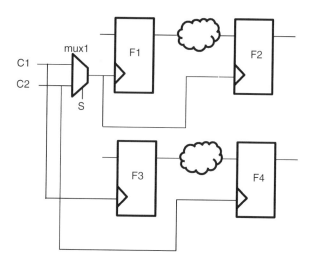

do interact for flops *F3* and *F4* and are only partially exclusive. In mode 2, clocks have the same period but the select line of the mux is set to *1*. Now in merged mode the select line cannot be set to a constant value. Thus, both clocks can pass through the mux. The interaction between *C1* and *C2* need not be timed for flops *F1* and *F2*, but still needs to be timed for *F3* and *F4*. So, combinational generated clocks need to be defined at the output of mux and these generated clocks would need to be physically exclusive as well. The merged mode constraint would look like:

Merged_Mode
create_clock -name C1 -period 10 [get_ports C1]
create_clock -name C2 -period 20 [get_ports C2]

*create_generated_clock -name GC1 -combinational *
-source [get_pins mux1/A] [get_pins mux1/Z]
*create_generated_clock -name GC2 -combinational *
-source [get_pins mux1/B] [get_pins mux1/Z] -add

*set_clock_group -physically_exclusive -group GC1 *
-group {GC2}

In summary, in the event of a conflict, constraints may have to be dropped or both constraints have to be considered and additional constraints may have to be added to model any exclusivity of paths that may have been present in the individual mode.

Sometimes, for simplification, additional pessimism can be introduced. However, in today's designs which try to exploit the last bit of performance, it is not desirable to add too much pessimism. Also any point of time, the merged mode cannot be any more optimistic than the original mode.

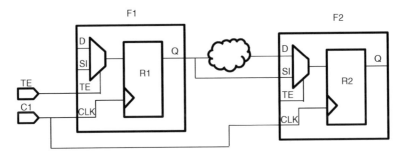

Fig. 15.8 Design with scan chain hooked up

15.4.4 Merging Functional and Test Mode

Let us consider the example in Fig. 15.8 where a scan flop has been used to hookup the design and the scan chain

Mode1
create_clock -name C1 -period 10 [get_ports C1]
set_case_analysis 0 [get_ports TE]

Mode2
create_clock -name C1 -period 40 [get_ports C1]
set_case_analysis 1 [get_ports TE]

In this example in functional mode the test enable pin (*TE*) is set to *0* and the clock has a period of *10ns*. In the test mode the test enable pin is set to *1* and the clock has a period of *40ns*. In the merged mode, both the clocks would have to be considered. However the *set_case_analysis* on the test enable pin would have to be dropped. In the merged mode this will enable paths from the *10ns* clock to all registers which are hooked to the Scan input pin (*SI*). Hence a false path would need to be defined from *10ns* clock to all the scan input pins (*SI*) for all flops in the design.

Merged_Mode
create_clock -name C1 -period 10 [get_ports C1]
create_clock -name C2 -period 40 [get_ports C1] -add
set_clock_group -physically_exclusive -group C1
-group C2
set_false_path -from C1 -to F*/SI

Let us understand the need to define *C2*, even though, the more restrictive clock (*C1*) has already been created. If *C2* was not defined then because of the false path the paths to *SI* are not getting timed. However they still need to be timed for the *40ns* clock. So, there is a need to define the additional clock *C2* on the same port. This also requires the clock group to be defined between *C1* and *C2*.

Alternately, instead of defining the clock C2, we could have used:

set_multicycle_path -to F*/SI -setup 4
set_multicycle_path -to F*/SI -hold 3

If you observe we have used a 4 cycle multi cycle path, since period of C2 is 4 times period of C1.

In the examples chosen, the clock names have been kept consistent; i.e., for similar clocks on similar objects, they have been given the same name in different modes. However, this is not a requirement. Often, the clock names could be different. Or, clocks with same name in two different modes could be having different characteristics or could be defined on different objects. Thus, all merge-related analysis and decisions should be based on the characteristics of the clocks, rather than on their textual names.

15.4.5 *Merging I/O Delays for Same Clock*

Let us see some examples of merging I/O delays.

Mode1
create_clock -name C1 -period 10 [get_ports C1]
set_input_delay -min 0.5 -clock C1 [get_ports P1]
set_input_delay -max 1.5 -clock C1 [get_ports P1]

Mode2
create_clock -name C1 -period 10 [get_ports C1]
set_input_delay -min 0.7 -clock C1 [get_ports P1]
set_input_delay -max 1.7 -clock C1 [get_ports P1]

In merged mode, the clocks would be merged according to the examples shown in aforementioned sections. In this example for sake of simplicity, let us assume the clocks are the same. To merge the input delays, pick the minimum of the delays specified using the *-min* option and the maximum of the delays specified using the *-max* option. The merged mode SDC would look like:

Merged Mode
create_clock -name C1 -period 10 [get_ports C1]
set_input_delay -min 0.5 -clock C1 [get_ports P1]
set_input_delay -max 1.7 -clock C1 [get_ports P1]

The same principle also applies to output delays.

15.4.6 *Merging I/O Delays with Different Clocks*

Let us take a variant of the example above. Let the clocks used to constrain the I/O delays be different in the two modes

Mode1
create_clock -name C1 -period 10 [get_ports C1]
create_clock -name C2 -period 20 [get_ports C2]
set_input_delay -min 0.5 -clock C1 [get_ports P1]
set_input_delay -max 1.5 -clock C1 [get_ports P1]

Mode 2
create_clock -name C1 -period 10 [get_ports C1]
create_clock -name C2 -period 20 [get_ports C2]
set_input_delay -min 0.7 -clock C2 [get_ports P1]
set_input_delay -max 1.7 -clock C2 [get_ports P1]

In merged mode, the input delay would need to be modeled with respect to both clocks and clock group would be required between the clocks. Since input delays are now defined with respect to two clocks the need for *-add_delay* option is required in the *set_input_delay* constraint.

Merged Mode
create_clock -name C1 -period 10 [get_ports C1]
create_clock -name C2 -period 20 [get_ports C2]
set_input_delay -min 0.5 -clock C1 [get_ports P1]
set_input_delay -max 1.5 -clock C1 [get_ports P1]
set_input_delay -min 0.7 -clock C2 [get_ports P1] \
-add_delay
set_input_delay -max 1.7 -clock C2 [get_ports P1] \
-add_delay
set_clock_group -logically_exclusive -group {C1} \
-group {C2}

The same principle also applies to output delays.

15.5 Challenges in Managing the Constraints

Irrespective of whether designers use Bottom-up or Top-down methodology with one mode or multiple modes, they face challenges in managing constraints. These include:

1. Blocks may meet timing when analyzed standalone, but when integrated at the chip level may fail timing. Typically, this happens if the block level constraints are not consistent with the constraints of other blocks, or with the top level constraints. For example, a block was signed off with a *10ns* clock, but at the chip level is being driven by *8ns* clock.
2. It may not be always possible to simply update the hierarchical name of the object to get the updated correct constraints. For example, a *create_clock* at block level may need to become a generated clock at chip level, if it is being driven by another block.
3. Critical paths which span block boundaries at chip level may not appear to be critical at block level. Typically, this happens if the allocation of delay across blocks and top level glue logic is not done correctly. For example, let us say, a path of *10ns* is spanning across three blocks. Each block is budgeted certain amount within this *10ns*. For one of the blocks, the number of logic levels is too

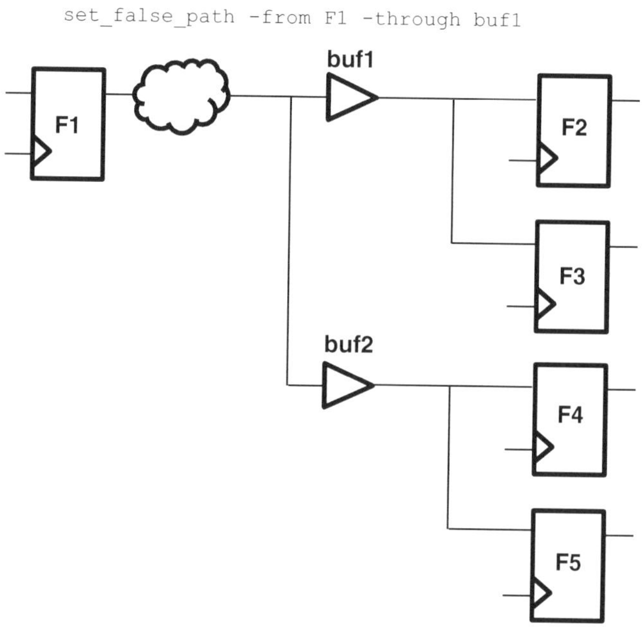

15.5 Challenges in Managing the Constraints

Fig. 15.9 Design with false path before optimization

low compared to the time being allocated to it. Thus, this block might not see this path as critical; while, in reality, this forms a segment of the critical path; any saving here would be useful at the top level.

4. Optimization done on a design may not mimic the constraints any more. Let us consider the example in Fig. 15.9. There is a false path set from the *F1* through buffer *buf1*, this means path from $F1 \rightarrow F2$, $F1 \rightarrow F3$ are false paths.

Now if design is optimized in a way the buffers are resized such that the same buffer drive all the flops, the equivalent design would look like Fig. 15.10.

However if the original constraints are not modified, then this optimization will result in additional paths ($F1 \rightarrow F4$, $F1 \rightarrow F5$) to be incorrectly constrained as false path. That could be a potential chip killer.

5. As described in Chap. 14, multimode analysis challenges include timing closure iterations and missed timing paths.
6. Mode merge sounds theoretically a strong concept but its actual success in the chip design has been limited because of its inherent limitation on what can be merged. In an attempt to remove conflicts, the addition of new constraints may cause unwanted optimization. Though not clear, mode merge may be leaving chip performance and area on that table, as it is a speed vs. accuracy tradeoff.

As of now, there are no algorithmic ways to merge modes for all kinds of scenarios. The examples shown above provide the general concept and cover most of the common scenarios encountered.

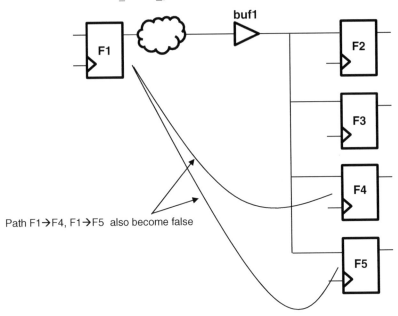

Fig. 15.10 Design after optimization

15.6 Conclusion

In summary, merging and propagating the constraints in your design flow is a tough problem. Designers use a combination of internal scripts, methodology; complemented by capabilities of commercially available tools to meet their current needs. There is active research and development going on in this area. As of now, most of these manipulations and merging are done based on individual experience.

Chapter 16
Miscellaneous SDC Commands

In this chapter, we will discuss some additional SDC commands and concepts.

16.1 Operating Condition

Delay of an element depends on several factors. The most important ones being

- Load on the element
- Input slew time
- Process
- Voltage
- Temperature

Out of the aforementioned factors, load and input slew are either computed for each element based on the design connectivity and properties available in the technology library or given as constraints, as explained in Chap. 10.

The information related to other three factors (Process, Voltage, and Temperature) is given through Operating Condition.

Conventional thoughts say that with higher temperature, usually the delay increases. Though, with current technology nodes, temperature inversion effects are also known to exist, where sometimes lower temperatures could have higher delay. With higher voltage, usually the delay decreases. Similarly, *P*-transistors with higher strength will cause transitions to *1* to be made faster; and *N*-transistors with higher strength will cause transitions to *0* to be made faster.

Temperatures can range from an extremely low value to a very high value. Voltage usually has a much smaller range. Process could range from Strong *P* or *N* to a Weak *P* or *N*. Usually, processes are characterized using different terminologies. Some of the popular terminologies include

- *Min/Typ/Max*: Indicating transistors with high strengths (i.e., min delay), moderate strengths or lowest strengths (i.e., max delay).
- *Slow/Typ/Fast*: Indicating transistors with weak strengths (hence, slow to drive), moderate strengths, or higher strengths (hence, fast to drive).
- *SNSP/SNWP/WNWP*, etc.: Indicating strengths of P and N transistors individually, e.g., *SNWP* means Strong *N*, Weak *P*. Thus, transitions to *0* will be faster, but transitions to *1* will be slower.

Between temperature, voltage, and process corners, there can be many variations or combinations. A user could explicitly specify the values of temperature, voltage, and process corner, under which analysis needs to be performed. SDC itself does not provide a set of commands to specify these parameters. Specific tools would have their own set of commands.

However, a more common practice is to define a set of operating conditions. Each operating condition is a combination of temperature, voltage, and process.

Different libraries could be using different naming conventions for operating conditions. One of the more common naming conventions used is *<speed><application>*, where

speed is one of

- *Best Case*: indicates fast conditions, such as low temperature, high voltage, strong transistors
- *Typical Case*: average conditions
- *Worst Case*: indicates slow conditions, such as high temperature, low voltage, weak transistors

Application is one of

- *Commercial*: very low range swing
- *Industrial*: slightly higher range swing
- *Military*: extremely high range swing

For example, *BCMIL* (Best Case Military) could be an operating condition and *WCIND* (Worst Case Industrial) could be another operating condition. For a set of operating conditions, using the above naming convention, the expected order in terms of fastest to slowest conditions would be

- *BCMIL*
- *BCIND*
- *BCCOM*
- *TYP*
- *WCCOM*
- *WCIND*
- *WCMIL*

If a user specifies an operating condition, he is effectively specifying a combination of temperature, voltage, and process corner. So, the timing analysis would be done under the specified set of environmental conditions.

16.1.1 Multiple Analysis Conditions

Sometimes, a user might want to perform different kinds of analysis under different conditions. Some of the conditions used more commonly for analysis are

- Slow (Worst Case/Max): All the paths will be assumed to have slowest delays.
- Fast (Best Case/Min): All the paths will be assumed to have fastest delays.
- Slow for Setup and Fast for Hold analysis: During setup analysis, all the paths will be considered to have slowest delays; while during hold analysis, all the paths will be considered to have fastest delays.
- On Chip Variation: This considers that there will be variations in operating conditions within the chip itself. Thus, different parts of the device will be using different kinds of delays. This is explained in Sect. 3.8 of the book.

All timing analysis tools might not necessarily support each of these analysis conditions. Or, some tools might even call these conditions by a different name. You might need to understand what all analysis modes are supported by your tool, and decide which one is most suited for your analysis.

16.1.2 set_operating_conditions

The SDC command to specify the operating conditions and the type of analysis to be done is

set_operating_conditions   [-*library* lib_name]
                            [-*analysis_type* analysis_type]
                            [-*max* max_condition]
                            [-*min* min_condition]
                            [-*max_library* max_lib]
                            [-*min_library* min_lib]
                            [-*object_list* objects]
                            [condition]

Analysis_type indicates the type of analysis that should be performed, whether to use slowest (max) or fastest (min) or On Chip Variation-based delay conditions.

Various Condition options are used to specify the operating point that should be considered for that condition, e.g., *max_condition* means the operating point that should be considered for slowest analysis. If the analysis mode is chosen to be a single condition (e.g., only slowest for all paths or only fastest for all paths), then just one operating point specification is sufficient.

The various library options are used to specify the libraries where the corresponding operating conditions are specified (in terms of temperature, voltage, and process).

Objects can be used to specify the design objects for which these operating conditions are to be used. In general, the operating condition is set for the whole design; hence, this is rarely specified. Earlier, this option was used to specify specific voltage values, if parts of the design were operating at a different voltage. However, now, the same purpose can be achieved through *set_voltage* command of SDC.

For example, the command: *set_operating_conditions -analysis_type min_max -max WCMIL -min BCIND -library L1* says

- Timing Analysis should be done under *min_max* condition, assuming, the tool has *min_max* as a valid analysis type. Lets say, in the given tool, the *min_max* analysis means, setup using Max conditions and hold using Min conditions.
- For Min conditions, use the operating condition as specified by BCIND.
- For Max conditions, use the operating condition as specified by WCMIL.
- The definition for the two operating conditions (BCIND and WCMIL) needs to be picked from the library *L1*.

16.1.3 Derating

Temperature, voltage, etc., are continuous variables. It is not possible to characterize delays for a library for all combinations of these variables. It is possible that the characterization data is not available for the operating condition chosen. In such cases, derating is used to obtain the values at the operating condition.

Let us say, values are measured at two different operating conditions where temperature is the only characteristic which is different among the two operating conditions. By knowing how much does the delay change due to a given change in temperature, it is possible to find out the scaling factor for delay as a function of temperature. This change in delay per unit change in temperature is called derating factor. If the delay value at a specific temperature is known, the delay at any nearby temperature can be obtained using the derating factor. This concept is called derating. The example provided is for delay derating due to temperature.

Thus, if the characterization data is not available for the operating point of interest, user can use the characterization data available for the nearest operating point and derate the values. The SDC command for specifying the derating is

```
set_timing_derate    [-cell_delay]
                     [-cell_check]
                     [-net_delay]
                     [-data]
                     [-clock]
                     [-early] [-late]
                     [-rise] [-fall]
                     derate_value
                     [object_list]
```

The derate_value is used to multiply the characterization data. Usually delay decreases for voltage increase. If the operating condition has a higher voltage compared to the characterization voltage, the derate_value would be less than 1; and so on.

The various options specify, whether the specified derate_value has to be used for

- *-cell_delay*: Cell delay
- *-cell_check*: Setup/Hold values for a cell
- *-net_delay*: Net delay
- *-data*: For data path
- *-clock*: For clock path
- *-early*: For early (hold) analysis
- *-late*: For late (setup) analysis
- *-rise*: For rise transitions
- *-fall*: For fall transitions

As can be seen, the command provides a very fine granular control over what do you want to derate and by how much.

However, you should be very careful, before derating any library data. Derating has to be applied based on characterization data. Usually, library data contains appropriate derating factors. You should be very sure if you want to override the derating factors provided in the library.

By specifying different derate_value for clock and data, the clock and data path can be given different derating factors. This is often used to mimic the on chip variation effect.

Earlier, max delay would happen at highest temperature, lowest voltage, and slowest process. However, due to temperature inversion effect, sometimes, highest temperature might not necessarily mean max delay. Also, the data path at max delay may not necessarily be the worst case scenario. Though conventionally, a lot of literature uses the term "worst case" for max delay.

16.2 Units

You might have noticed that in most of the examples mentioned for the SDC commands, the numbers do not have a unit. In the textual explanation of the examples or commands, we have used *ns*, only for the sake of convenience.

Units are picked up from the library being used. SDC does not have the concept of units being associated with each command. It allows specifying units in a command by itself. The SDC command for specifying units is

set_units   *-capacitance* cap_unit
            -resistance res_unit
            -time time_unit
            -voltage voltage_unit
            -current current_unit
            -power power_unit

This command can be used to specify the units for capacitance (e.g., used in *set_load*), resistance (e.g., used in *set_resistance*), time (e.g., used in *set_input_delay*), voltage (e.g., used in *set_voltage*), current (to derive units for power – in conjunction with unit for voltage), and power (e.g., used in power optimization related commands).

The standard itself does not specify how to interpret the *set_units* command. At the time of writing the book, for most tools, this command does not change the unit. It only acts as a documentation of user's intent – as to what unit did he intend when specifying all the numbers in the SDC commands. Tools would check these units against the units that they are using (from the library), and flag if there is a mismatch in units being used by the tool (from library) and being used by the user (specified with *set_units*).

16.3 Hierarchy Separator

Most electronic design languages allow a user to refer to a hierarchical object, by providing the concept of a hierarchical name. In all such languages, a change in hierarchical boundary is denoted by a specific character, e.g., in Verilog, "/" denotes a hierarchical boundary (popularly called, hierarchy delimiter or hierarchy separator). Sometimes, a user might want to put this same delimiter character as a part of the name itself. In such situations, the languages allow for an escape character, so that the hierarchy delimiter character might no longer be treated as a hierarchy delimiter, rather it should get treated as an ordinary character, which is a part of the name. Again, taking the example of Verilog, the escape character is "\". Thus, in Verilog, " *a/b* " means object "*b*" within instance "*a*". However, " *a/b* " in Verilog means an object, whose name contains the character "/".

SDC does not provide the concept of an escape character. A reference to "*a/b*" in SDC could mean either of

- "*b* within *a*" (i.e., " *a/b* " of Verilog) or
- "/" being part of the name itself (i.e., " *a/b* " of Verilog).

This could result in ambiguous identification of an object. In order to remove this ambiguity, SDC allows a user to specify his own hierarchy separator.

16.3.1 set_hierarchy_separator

The SDC command for specifying hierarchy separator is

set_hierarchy_separator separator

You should choose your hierarchy separator as that character, which is not used in the actual naming of any object in your design.

16.3 Hierarchy Separator

Let us say, your design does not use any escaped naming. In that case, "/" is a good hierarchy separator. When SDC refers to " a/b ", there is no ambiguity; it is clear that this mean "b within a". On the other hand, let us say, your design has several objects, where "/" is used as a part of the escaped name itself. Using this same character as a hierarchy separator can cause ambiguity – as mentioned in the previous section. In such situations, the hierarchy separator might be changed to another character (say: @) using the following command:

set_hierarchy_separator @

With the hierarchy separator being set at @, the above ambiguity is resolved. " a/b " only means the design object " $\backslash a/b$ ", where "/" is part of the name. It cannot match "b within a". If we want to refer to "b within a", we need to specify $a@b$.

16.3.2 -hsc

Besides setting a global hierarchy separator, SDC also allows to set a hierarchy separator in a very local context, where the scope of the specified separator is limited only to objects specified within a specific command.

This capability is useful in the following two situations:

Let us say, your design uses too many escaped names, with all kinds of characters in them. So, irrespective of which character do you choose for hierarchy separator in SDC, some object in your design might already be using that same character. In such situations, you might want to use a different hierarchy separator at different places in the SDC.

Usually, readability is an important aspect of an SDC file. As seen in the previous section, if we have used "/" as part of a name, we have to choose a different hierarchy separator (say: @ – as shown as an example in the previous section). Because of this, all the design object references might appear funny (using "@", rather than "/", which a user might be more used to). Let us say that there are a very few objects which have such names which may cause ambiguity. In such a case, we can decide to use a different hierarchy separator – only at commands, where there are chances of ambiguity.

The way to specify a hierarchy separator with a local scope is to add the option *-hsc*.

The following command could mean the start point to be "b within a" or " $\backslash a/b$ ":

set_false_path -from [get_pins a/b]

However, the following command only means the pin " $\backslash a/b$ ", where "/" is part of the name itself:

set_false_path -from [get_pins a/b -hsc "@"]

If hierarchy separator has to be made different for a lot of commands, it is better to use the *set_hierarchy_separator* command, which has a global scope. On the other hand, *-hsc* option might be preferred, if the hierarchy separator has to be changed for only a few objects.

16.4 Scope of Design

The scope of working design is specified through SDC command *current_design*. All objects referred in subsequent commands are with respect to the *current_design*. The *current_design* is the effective "top" of the design. Anything outside the hierarchy of *current_design* is not accessible. Thus, *current_design* does not have any instance name for itself.

It should be noted that even though, these commands are part of SDC, but several tools expect these commands to be specified in their shell or project settings, rather than being read through SDC file itself.

16.4.1 current_instance

The scope can be changed to an instance within the *current_design*. The SDC command to limit the scope to a specific instance is

current_instance [instance]

When the scope is changed to a specific instance, all search and query is limited to within the specific instance. So, search and access commands can be specified using names relative to the *current_instance*. However, names returned are still always with respect to the current_design.

Let us consider the design shown in Fig. 16.1.

If we search for *U2* (through: *get_cells U2*), we will get the topmost *U2*, instantiated directly under the "top".

However, if we want to get *U2* under *U1*, there are two ways of getting access to it.

get_cells U1/U2

OR

current_instance U1
get_cells U2

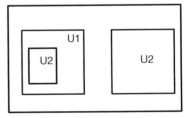

Fig. 16.1 Object reference through current_instance

In the second example, the search for *U2* is made only inside *U1*, and it returns back *U1/U2*. Notice that though the search was limited to be only within *U1*, the return name is with respect to the *current_design*, rather than the *current_instance*.

Setting up *current_instance* is useful, when you want to limit the scope of your search to within a specific portion of the hierarchy. Search based on *current_instance* might be more efficient, compared to searching through the entire design and then using the TCL filters to select only those objects whose initial portion matches the desired hierarchy.

16.5 Wire Load Models

We've seen earlier that delay depends on capacitive load. Nets also provide a significant capacitive load. The net capacitance in turn is dependent on the length of the specific net. Before routing is done, the actual net lengths, etc., are not known. During pre-route stage, tools estimate the net lengths, and use this estimated lengths to compute the wire-capacitance.

Wire load models are used to provide a statistical estimate of the wire-lengths. Usually, wire-lengths depend on

- Size of the design: A larger design will typically have longer wires, as they have to span across larger size.
- Fanout: A higher fanout will typically mean longer wire-lengths, as more pins need to be connected.

Wire length estimates (hence, wire load models) are useful only for pre-layout stage. Once routing is done, the tools have access to the actual net length and those get used, rather than the wire load models.

The wire load models can be set through the SDC command:

set_wire_load_model   -*name* model_name
[-*library* lib_name]
[-*min*] [-*max*]
[object_list]

The specified model name is searched for in the given library. -*min* and -*max* are used to specify, whether the specific model is to be used for min conditions or for max conditions. Typically, users don't change wire load models for analysis conditions. *Object_list* is also not used very often. A more popular mechanism is to change the scope of the design and then apply the wire load model on that scope, without having to specify the object_list explicitly.

So, *set_wire_load_model* -*name* WIRE_LOAD_70X70 indicates that for the given scope, the tool should use a *wire_load_model* named *WIRE_LOAD_70X70*. The actual model would be defined in the library loaded into the analysis tool. Usually, the model provides wire length estimates as a function of fanout.

In most cases, the user would provide a wire load model based on the size of the design. For example:

current_design Top
set_wire_load_model -name WIRE_LOAD_100X100

current_instance Next_level
set_wire_load_model -name WIRE_LOAD_70X70

current_instance Lowest_level
set_wire_load_model -name WIRE_LOAD_40X40

Hence, most tools provide a mechanism to automatically assign the *wire_load_model* appropriate for a given size, rather than the user having to specify different wire load models for different portions of the design.

16.5.1 Minimal Size for Wire Load

Let us say, that you have set the tool for automatic assignment of wire load models. However, you might want that for the purpose of wire load model selection, the block size should always be considered to be larger than some specified value. This can be specified through the SDC command:

set_wire_load_min_block_size size

For any block smaller than the specified size, the wire load model would be assigned, as if the block had an area of the specified size. Say, you want that the minimum size that should be considered is *30*, so you can specify

set_wire_load_min_block_size 30

If any block has an area smaller than *30*, the area would still be considered as *30* for wire-load estimation. For blocks larger than *30*, their actual area would be used.

16.5.2 Wire Load Mode

Let us consider the wire *n1* in Fig. 16.2.

In order to estimate the wire-length, which instance's area should we consider? Should it be for *M2*, from which the net originates? Or, should it be *M3* or *M4* into which the net feeds in? Or, should it be *M1* within which most of the net lies? Or, should it be *M0*, which fully encompasses the net? Or, should it be the top level?

This selection can be made through the SDC command:

set_wire_load_mode mode_name

16.5 Wire Load Models

Fig. 16.2 Wire spanning across multiple hierarchies

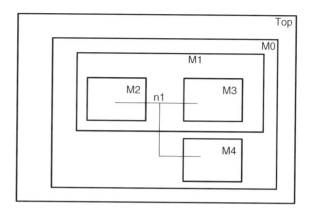

The mode_name specified should be one of the modes that your analysis tool supports. Some of the more commonly supported modes are the following:

- Use the top level for the whole design – including for nets within the sub-hierarchies.
- Use the smallest module which fully encloses the whole net.
- Use different instances for different segments of the net. For example, different models (appropriate for the respective sizes) to compute the segment lengths for portions within *M2, M3, M1, M4,* and *M0.* Finally, add up the lengths in each segment to decide on the total length.

You have to find the modes that your tool supports and identify which of these modes you prefer for your analysis.

16.5.3 Wire Load Selection Group

Many libraries define collection of wire loads which may be applicable to different cells based on the area. Let us consider an example extract from a Synopsys .lib file

wire_load_selection (WL_SELECTION_GRP_1) {
  wire_load_from_area (0, 4000, "WIRE_LOAD_20X20");
  wire_load_from_area (4000, 49000, "WIRE_LOAD_70X70");
  wire_load_from_area (49000, 100000, "WIRE_LOAD_100x100");
}

A library may contain many such selection groups. The selection group of interest can be assigned using the *set_wire_load_selection_group* command. The syntax for this command is

set_wire_load_selection_group   [*-library* lib_name]
                                               [*-min*] [*-max*]
                                               group_name
                                               [object_list]

The *-min and -max* command imply the selection group that needs to be picked for minimum and maximum conditions. The *-library* indicates the name of the library from where the collection group specified using "group_name" needs to be picked up. Generally the collection group applies to the *current_instance*. However the user can apply the collection group to a specific hierarchical cell, though this option is not used that often.

16.6 Area Constraints

These constraints are not useful for timing analysis. However, these are part of SDC, because they impact implementation tools like synthesis, in the sense that they impact their ability to optimize on area.

Let us consider a situation, where your top level design is made of various IPs. These IPs are generic IPs which provide customization, based on the values on some of their input pins. Let us say, in your specific design, some of these pins are to be hard-wired to a specific value (say: *0*). In this case, synthesis tool should be allowed to optimize the logic driven by this pin – so that it is good enough for logic *0* only.

Let us further say, you are using a bottom-up methodology, wherein this IP is being synthesized by itself. Since the connections to this IP are not visible at this stage, synthesis tool does not know, that there are pins of this IP which will be hard-wired during instantiations to logic *0*. This information can be conveyed to the tool, through the SDC command:

set_logic_zero port_list

Synthesis tools would assume the specified ports to be hard-wired to *0* and will optimize the logic inside the IP – in order to reduce area. Similarly, SDC allows

set_logic_one port_list

to convey that the specified ports are to be considered as hard-wired to *1*.

set_logic_dc port_list

to convey that the specified ports are dont care. The logic driven by these ports can be reduced significantly by the synthesis tools.

It should be noted that these commands are different from *set_case_analysis* discussed in Chap. 14. *set_case_analysis* sets the logic to a specific value only for a specific analysis. It is still understood that the opposite logic value will also be a valid value on that object, and functionality cannot be optimized based on *set_case_analysis*. It effects only timing portion of the analysis, while *set_logic* commands impact the functionality itself.

Synthesis tools can also be provided an area target. The SDC command for specifying an area target is

set_max_area area_value

16.7 Power Constraints

The synthesis tool will try to realize the design in a manner such that the area finally required is lesser than the specified area value. It should be understood that area and performance (and power) are often mutually contradictory goals. For a given technology node, if area improves, performance and/or power will likely deteriorate. So, while specifying the constraints, we should be realistic, rather than trying to stretch the tool too much in any one dimension, as it might deteriorate the other two dimensions.

16.7 Power Constraints

Because of power considerations, it is common in today's designs to have different parts of the design working at different voltage levels. This brings in its own set of challenges. It is beyond the scope of this book to explain power-related concepts. However, in this section, we will briefly touch the SDC constraints which come into effect due to power or multi-voltage situations.

We've already discussed that the delay is dependent on voltage. If a specific portion of a device is operating at a different voltage (compared to the operating condition specified), its delay would be different. The SDC command to specify which portions of the device are operating at a different voltage is

set_voltage   [-*min* min_case_value]
         [-*object_list* list_of_power_nets]
         max_case_voltage

max_case_voltage refers to the voltage corresponding to the maximum delay (viz., lowest voltage). -*min* is used to specify the voltage corresponding to the minimum delay (viz., highest voltage). Note that the -*min/max* is with respect to the delay, rather than the voltage!

-*object_list* specifies the list of power_nets which are at the specified voltage.

The next set of commands does not directly impact timing analysis. However, they impact several implementation tools. Timing would be a side-effect of the layout/routing changes.

16.7.1 Voltage Island

If a design has multiple power supplies, it makes sense to localize all devices operating at a given voltage in one area. Such an arrangement makes it easy for the power rail distribution, as all the rails don't have to go all over the chip. Specific rails will go to specific locations, and all cells connected to that rail should be lying in that location. This concept is also called voltage island.

The SDC command to specify voltage islands (so that placement tools can place cells accordingly) is

create_voltage_area -*name* name
[-*coordinates* coordinate_list]
[-*guard_band_x* float]
[-*guard_band_y* float]
cell_list

Name specifies a name given to the specific voltage island. cell_list is the list of cells, which needs to be placed together. This could even be orthogonal to hierarchy; meaning: cells belonging to two different hierarchies could be in the same voltage island; similarly, something lower in the hierarchy could be in a different voltage island compared to where it is instantiated.

-*coordinates* specify the rectangular region within which the specific voltage island should be kept. -*guard_band_x* and -*guard_band_y* specify a distance along x-axis and y-axis, where no cells should be placed. This is the buffer area between two different voltage islands.

16.7.2 Level Shifters

When a signal moves from one voltage level to another, its noise margin gets impacted. In order to ensure a clean transfer of signal across voltage domains, level shifters are often used. Synthesis tool would need to know when to insert level shifters. The SDC commands to convey this direction are

set_level_shifter_strategy [-*rule* rule_type]

set_level_shifter_threshold [-*voltage* float] [-*percent* float]

Depending upon your choice of methodology, your strategy could be one of the following:

- Put level shifters, whenever a signal goes from one voltage to another.
- Put level shifters, only when a signal goes from a lower voltage to a higher voltage.
- Put level shifters, only when a signal goes from a higher voltage to a lower voltage.

You would need to check your power optimization tool to see which of these strategies it supports and accordingly specify the rule name.

Also, you might not want to put level shifters for even a minor change in voltage levels. The *set_level_threshold* command allows you to control the variation in voltage levels, when the level shifter should be put in. Only when the voltage variation is higher than the specified value, does the optimizer put in the level shifter. The variation can be expressed in absolute voltage diff or in terms of % diff.

16.7.3 Power Targets

As mentioned in Sect. 16.6, area, power, and timing are mutually contradictory. A user can specify power requirements also to an optimization tool, so that it tries to keep the power values also within required limits. The SDC commands for specifying power targets are

set_max_dynamic_power power [unit]
set_max_leakage_power power [unit]

These commands set the upper target for dynamic and leakage power, respectively. The synthesizer is expected to keep the power values to be below the specified limits.

Power has become an important consideration all by itself. There are dedicated languages (UPF and CPF) to specify power intent. The importance of SDC commands for power is getting reduced.

16.8 Conclusion

The commands mentioned in this chapter provide the operating environment for the design. SDC commands are used by tools other than timing analysis. For implementation tools, they specify area and power targets also. Usually, area, power, and timing performance are mutually contradictory. Hence, area and power constraints impact final timing performance also.

SDC has some more commands also. With the understanding gained so far, it should be fairly simple for a reader to understand and interpret other commands and their usage scenarios.

Some specific tools have extended SDC with some additional proprietary commands or options. These extensions allow these tools to get additional user input for something that is specific to those tools only. In the next chapter, we will see proprietary extensions for current generation of Xilinx tools.

Chapter 17
XDC: Xilinx Extensions to SDC

Frederic Revenu
Xilinx
frederic.revenu@xilinx.com

FPGA design flows have become very similar to ASIC flow. They both usually start from RTL and require a number of similar physical and timing constraints in order to ensure proper functionality and timing on hardware. Adoption of industry standards has also helped with the convergence between the two worlds, particularly on the timing constraints side via the adoption of SDC and deprecation of proprietary equivalent formats. The main differences remain around the rules to be followed during the design implementation. ASICs come with an extensive set of manufacturability and testability rules, while FPGA designs need to follow a set of higher level rules such as device capacity and architecture features compatibility. These differences are reflected in the FPGA design flow where the tools are able to simplify or hide a number of complex rules that are typically encountered in ASIC flows, for example on signal integrity, and automatically create some constraints such as generated clocks or jitter. The following chapter will focus on how SDC support has been extended in Xilinx new generation of FPGA compilation software. The Xilinx extension to SDC is called *XDC*.

17.1 Clocks

Compared to ASICs, FPGA devices provide a limited amount of dedicated clocking resources with known characteristics which should be used as much as possible by any design. These resources are a combination of input buffers, clock buffers, clock-modifying blocks such as PLL and MMCM (Mixed-Mode Clock Manager), and clock recovery or generation blocks such as Gigabit Transceiver and PCIe. The clock signals are distributed with dedicated routing resources which span across a group of IO ports, a portion of the device or the full device. While the designer remains in charge of defining the clocks coming from outside the device, the Xilinx FPGA compilation software is able to complete and refine the definition of these clocks as they propagate through the design.

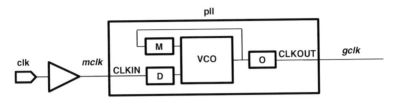

Fig. 17.1 Tool-created generated clock on a Xilinx PLL output

17.1.1 Primary and Virtual Clocks

Like in a standard ASIC flow, the primary and virtual clocks need to be defined with the *create_clock* command as presented in Chap. 5.

17.1.2 Generated Clocks

There are two categories of generated clocks in Xilinx FPGA compilation software. They are

- User-defined generated clocks (as explained in Chap. 6)
- Tool-created generated clocks

17.1.2.1 Tool-Created Generated Clocks

When a clock traverses a Xilinx-specific clock-modifying cell, the tool knows the characteristics of the clock(s) at the output of that same cell and automatically defines the corresponding generated clocks. The clock transformation is fully described by the cell parameters.

Let us consider the example usage of PLL shown in Fig. 17.1. Let the reference clock have a period of *10 ns* with a *50 %* duty cycle:

create_clock-name mclk-period 10-waveform {0 5} [get_ports clk]

Let us further assume that the PLL instance has the following parameters:

- Clock input divider: $D=2$
- Feedback clock multiplier: $M=32$
- Output clock divider: $O=8$

(The divider and the multiplier is in terms of frequency and not period. For period computations, divider becomes multiplier and vice-versa).

The period of the generated clock at the output of this PLL is given by

17.1 Clocks

$$Fgclk = \frac{M}{D*O} * Fmclk \quad (17.1)$$

The equivalent constraint automatically created by the tool is

*create_generated_clock -name gclk *
*-edges {1 2 3} -edge_shift {0 -2.5 -5} *
-source [get_pins pll/CLKIN] [get_pins pll/CLKOUT]

The name of the generated clock is based on the name of the net directly connected to the PLL output pin, at the same level of hierarchy, i.e., *gclk* in this particular example.

If several automatically generated clocks have a name conflict, the timing engine will append a unique index to each name. For example, *gclk_0*, *gclk_1*, etc.

Since the generated clock is automatically created, and its waveform is based on the actual behavior of the PLL's parameter, the user does not have to explicitly put a *create_generated_clock*. A user may still decide to explicitly define a generated clock, if (for example), he wants to give a specific name to this clock. If the user has already defined a clock on the output pin of the clock-modifying block, the corresponding auto-generated clock will not be created. This ensures that the user still has full control over the clock definitions.

17.1.2.2 Generated Clocks with Nonintegral Ratio

The Xilinx devices provide a large number of PLL and MMCM to be used in diverse situations: clock insertion delay compensation, frequency synthesis, jitter filtering. In many cases, the clocks generated by these blocks have a frequency ratio described by (17.1) (in Sect. 17.1.2.1). It is possible for the user to describe this ratio by using the *create_generated_clock* options *-multiply_by* and *-divide_by* simultaneously.

Considering another example, with a different set of parameter values (as given below), the *create_generated_clock* constraint can be written as follow:

- Clock input divider: $D=3$
- Feedback clock divider: $M=32$
- Output clock divider: $O=7$

*create_generated_clock-name gclk-multiply_by 32-divide_by 21 *
-source [get_pins pll/CLKIN] [get_pins pll/CLKOUT]

The advantage of using *-multiply_by* and *-divide_by* options is that the values purely describe the clock transformation through the cell and are independent of the master clock waveform definition. Hence, they don't need to be updated if the master clock definition gets changed. On the other hand, -edge/edge_shift depends on the exact master clock waveform.

Note: the simultaneous use of *-multiply_by* and *-divide_by* is not compatible with the introduction of a phase shift in the generated clock definition.

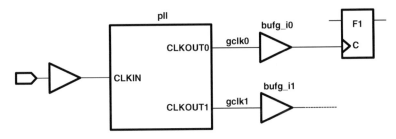

Fig. 17.2 Getting a clock along its tree

17.1.3 Querying Clocks

In general, the name of the tool-created generated clocks is predictable and can safely be used in other timing constraints. But in some cases, for the reason explained in Sect. 17.1.2.1, this may not be true. Usually, clocks are queried by using the *get_clocks* command, where the exact clock name or a name pattern must be supplied. This use model has some limitations in case the names are not known. For this reason, Xilinx tools support two unique additional *get_clocks* switches:

17.1.3.1 -of_objects

This option allows querying the design database and retrieving objects related to other ones based on their relationship. *get_clocks -of_objects* returns the clock objects which traverse the specified pins, ports, or nets. It does not work with cell objects, and does not return anything for objects which do not belong to a clock tree.

For the circuit shown in Fig. 17.2, a primary clock *clk* traverses a PLL instance, which generates two clocks, *gclk0* and *gclk1*. These two clocks propagate through their respective clock network, i.e., clock buffers *bufg_i0* and *bufg_i1*, and reach the sequential cells. The *get_clocks -of_object* command can be used on various objects as follows:

get_clocks -of [get_pins pll/CLKOUT0]
get_clocks -of [get_nets gclk0]
get_clocks -of [get_pins bufg_i0/O]
get_clocks -of_objects [get_pins F1/C]

All the commands above return the same result: *gclk0*.

This option can be used similarly with some other get commands, such as *get_pins*, *get_nets*, *get_cells*, and *get_ports*.

17.1 Clocks

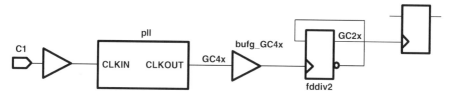

Fig. 17.3 Getting a clock and its associated generated clocks

17.1.3.2 -include_generated_clocks

A master clock and its associated generated clocks define a group of synchronous clocks, i.e., clocks with a predictable initial phase relationship. The *get_clocks -include_generated_clocks* command returns a list of clocks which includes the specified master clock(s) as well as the associated generated clocks, plus their child generated clocks if any. This is particularly convenient for creating asynchronous clock groups constraints (see Clock Groups described in Sect. 17.1.4).

For the circuit shown in Fig. 17.3, the primary clock *C1* reaches a PLL instance, *pll*, which generates the *GC4x* clock. One of the GC4x clock tree branches reaches a register-based clock divider, *fddiv2*, which in turn generates the clock *GC2x*. The corresponding clock definitions are presented below, as well as the result of *get_ clocks -include_generated_clocks*.

create_clock -name C1 -period 10 [get_ports C1]
*create_generated_clock -name GC4x -multiply_by 4 *
 -source [get_ports C1] [get_pins pll/CLKOUT]
create_generated_clock -name GC2x -divide_by 2
 -source [get_pins pll/CLKOUT] [fddiv2/Q]

get_clocks -include_generated_clocks C1
will return: *C1 GC4x GC2x*

17.1.4 Clock Groups

Section 7.3 of the book explains how the various options of *set_clock_groups* are important only for signal integrity analysis. From timing analysis perspective, the options are interchangeable. At the beginning of this chapter, we mentioned that the FPGA designers do not have to handle signal integrity issues the same way as with ASICs. This is mostly because of the way the internal hardware circuitry is designed and how the cell timing arcs and net delays are calculated.

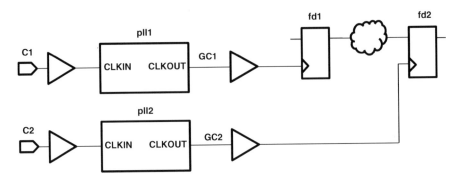

Fig. 17.4 Asynchronous clocks

In order for the FPGA device to safely operate under a wide range of operating conditions, the timing models include higher margin. Thus, for Xilinx users, the *set_clock_groups* options related to crosstalk are not needed and can be used interchangeably

- *-logically_exclusive*, *-physically_exclusive*, and *-asynchronous* are all treated similarly by the timing engine. They can still be used by the designer in order to clarify the original intent of the *set_clock_groups* constraint and maintain compatibility with ASIC flows.
- *-allow_paths* is completely ignored.

The example below shows how to simply specify that two groups of clocks are asynchronous based on their topology, and without knowing the name of the clocks. The schematics presented in Fig. 17.4 above shows two primary clocks *C1* and *C2*, defined on input ports and propagating through PLL instances before reaching the sequential elements *fd1* and *fd2*. The logic path between *fd1* and *fd2* is assumed to be safe by design and should be ignored during the timing analysis.

create_clock -name C1 -period 10.57 [get_ports clk1]
create_clock -name C2 -period 6.667 [get_ports clk2]
*set_clock_groups -asynchronous -name asynch_clk1_clk2 *
    *-group [get_clocks -include_generated_clocks -of [get_ports C1]] *
    -group [get_clocks -include_generated_clocks -of [get_ports C2]]

Assume that the generated clocks are named *GC1* and *GC2*. In addition to the relationship between *C1* and *C2* as explained in Chap. 7, the following relationships have also been inferred due to *-include_generated_clocks*:

1. *C1* and *GC2* are also asynchronous to each other
2. *GC1* is also asynchronous to *C2* and *GC2*

17.1.5 Propagated Clocks and Latency

The Xilinx FPGA timing models are based on accurate delay extraction from the device hardware and are used throughout the implementation flow to provide the user with the most accurate estimated or actual routed net delays. This is particularly true for the clock distribution network. As explained previously in this chapter, the Xilinx FPGA architecture includes a dedicated and optimized clocking infrastructure which the designer should utilize as much as possible. A particular design clock tree propagates through certain clock buffer primitives, which indicates how far it will propagate and what route it is likely to take. This gives useful information to the Xilinx implementation tools for providing good clock insertion delay estimates before placement, and very accurate ones after placement. Thus, Xilinx FPGA tools consider all clocks as propagated by default in order to let the tools calculate the insertion delay for each of them and have a more accurate delay computation early in the design cycle. Consequently, the *set_propagated_clock* command is supported (for the sake of completeness) but not needed.

Also, it is not recommended for the user to manually set the clocks network latency as the Xilinx software can infer a better estimate, corresponding to the specific device size and speed grade. The user is still responsible for specifying the clocks source latency, via the *set_clock_latency -source* command, especially for clocks coming from outside the FPGA (see Sects. 8.5 and 8.6).

17.1.6 Clock Uncertainty

In addition to automatically provide the clock network latency, the Xilinx software computes the total clock uncertainty of a timing path as a function of the following variables:

- *System Jitter*: This is the uncertainty common to all clocks of the design, due to power and board noises. A default value is provided by the Xilinx software for each FPGA device family and can be overridden by the user via the *set_system_jitter* constraint.
- *Input Jitter*: This is the uncertainty of a particular clock at its definition point. It usually reflects the random, peak-to-peak jitter present on a clock generated outside the device, when it reaches the FPGA device package pin. The input jitter of a clock generated by an MMCM (or PLL) is derived from its master clock input jitter, based on the MMCM (or PLL) operating mode. A default value is provided by the Xilinx software for each FPGA device family and can be overridden by the user via the *set_input_jitter* constraint.
- *Discrete Jitter*: This is the jitter introduced by an MMCM (or PLL) instance only on the clocks it generates depending on its operating mode. Since these are hard blocks on the FPGA device, hence, Xilinx software knows an accurate value of this jitter. Thus, this value is always provided by the Xilinx software and cannot be modified by the user.

- *Phase Error*: This is the undesired phase variation introduced by an MMCM (or PLL) instance between two clocks it generates. This value is also always provided by the Xilinx software and cannot be modified by the user.
- *User Uncertainty*: This is the additional uncertainty specified by the user on a particular clock, or between two clocks, via the *set_clock_uncertainty* constraint (explained in Sect. 8.4).

The effective uncertainty that Xilinx software considers after combining the effect of various jitters, phase error and user uncertainty is given by the following equation:

$$\text{Timing Path Clock Uncertainty} = \frac{\sqrt{TSJ^2 + TIJ^2} + DJ}{2} + PE + UU \quad (17.2)$$

where,

- TSJ = Total System Jitter
- TIJ = Total Input Jitter
- DJ = Discrete Jitter
- PE = Phase Error
- UU = User Uncertainty

Total System Jitter corresponds to the quadratic sum of the launch clock system jitter and the capture clock system jitter. Since both clocks are propagated by default and the system jitter value is the same for all clocks, the Total System Jitter is usually

$$\text{Total System Jitter} = \text{System Jitter} * \sqrt{2} \quad (17.3)$$

Similarly, the Total Input Jitter corresponds to the quadratic sum of the launch clock input jitter and the capture clock input jitter

$$\text{Total Input Jitter} = \sqrt{\text{LC Input Jitter}^2 + \text{CC Input Jitter}^2} \quad (17.4)$$

where LC is the Launch Clock and CC is the Capture Clock.

In summary, by default, all clocks are treated as propagated clocks and the Xilinx tools provide an accurate estimate of the clock uncertainty. The user can add additional margin to his design by using the *set_clock_uncertainty* constraint.

17.2 Timing Exceptions

The Xilinx compilation software supports all timing exceptions as defined in the SDC standard. Only one option has been added to the max delay constraint: *set_max_delay -datapath_only*. When this option is used, the launch and capture clock insertion delays are removed from the slack computations. This is particularly convenient when constraining the maximum delay between two asynchronous clock domains where the designer wants to keep the latency as small as possible. In traditional ASIC flows, the designer has to consider the clock skew into the max delay value specified with the *set_max_delay* constraint. In Xilinx FPGA designs, third party RTL IPs are often used. They all include a section of clock tree which can look different once the top-level design is complete. By using the *datapath_only* option, the IP designers can define delays on clock domain crossing paths included in their IP, which are invariant of the clock skews.

When using the *-datapath_only* option, the min delay analysis is disabled on the same paths.

17.3 Placement Constraints

For complex designs or designs with challenging timing requirements, it is often necessary to guide the implementation algorithms by specifying physical constraints. The Xilinx compilation software offers several options:

- *Cell Placement*: The user can constrain the location of a cell by setting its *LOC* property. Following is an example of how to define a placement constraint on the LUT instance inst123:

    set_property LOC SLICE_X0Y0 [get_cells inst123]

    Depending on the primitive of the cell, it can be placed on a particular grid. For example, LUTs, DSPs, and RAMBs are placed on their specific grids. These grids have a different range for each device size or family, so any placement constraint is only valid for a particular device.

- *Net Weight*: The user can increase the importance of a net by setting its weight property to a higher value (the default value is *1*). The cells connected to the net with high weight are more likely to be placed closer to each other. This is particularly convenient for improving the placement of a group of cells without modifying their timing constraints or restricting them to a specific area. Following is an example of how to set a higher weight on the net *n456*:

    set_property WEIGHT 10 [get_nets n456]

- *Physical Block (pblock)*: Traditional design floorplanning is done by using *pblocks*, which is a convenient mechanism for keeping critical logic grouped together or clock to special hardware resources such as IO buffers for example. A *pblock* is a location range constraint set on a group of cells. The range is defined by one or more rectangles on the device floorplan. The following example shows how to create a pblock, add the hierarchical cell usbInst to it, and place it:

create_pblock pblock_usb
add_cells_to_pblock pblock_usb [get_cells usbInst]
resize_pblock pblock_usb -add {SLICE_X12Y34:SLICE_X56Y78}

Assigning a hierarchical cell to a pblock will assign all its children cells too.

17.4 SDC Integration in Xilinx Tcl Shell

Tcl is commonly used across most EDA tools as the user interface scripting language. Xilinx tools provide a convenient support of SDC and netlist objects which fully relies on the standard TCL built-in commands and on object substitution with their name depending on the context.

Tcl basics are covered in Chap. 4. The get_* commands in Xilinx tools return the string name for the objects, as prescribed by Tcl. This is different from some other Tcl-based tools, which provide a handle to the object, rather than its string name.

17.5 Conclusion

This chapter was used mostly to explain, how specific tools could extend SDC, in order to fit their specific requirements. These modifications could be in the form of

- additional switches (e.g., *-datapath_only*),
- additional commands (e.g., *set_property*),
- default application of certain constraints (e.g., *create_generated_clock*),
- etc.

For actual features of the Xilinx FPGA devices and Xilinx tools, including current extensions to SDC, you should refer to http://www.xilinx.com.

Bibliography

1. Using the Synopsys Design Constraints Format: Application Note, Version 1.9, Synopsys Inc., December 2010
2. Liberty User Guides and Reference Manual Suite, Version 2011.09, Open Source Liberty (www.opensourceliberty.org)
3. Churiwala S, Garg S (2011) Principles of VLSI RTL Design: A Practical Guide. Springer, New York.
4. Bhatnagar H (1999) Advanced ASIC Chip Synthesis: Using Synopsys® Design Compiler™ and PrimeTime®. Kluwer Academic, Boston
5. Keating M, Bricaud P (1999) Reuse Methodology Manual. Kluwer Academic, Norwell, MA
6. Sebastian Smith MJ (1997) Application-Specific Integrated Circuits. Addison-Wesley, Boston, MA
7. Bhasker J, Chadha R (2009) Static Timing Analysis for Nanometer Designs: A Practical Approach. Springer, New York

Index

A
active edge, 47
add, 50, 52, 66
add_cells_to_pblock, 217
add_delay, 104, 108
aggressor, 76, 79
all_clocks, 45
all_inputs, 45
all_outputs, 45
allow_paths, 79, 214
all_registers, 45
Application-specific integrated circuit (ASIC), 1, 4, 5, 209
arc, 123
area, 23, 35, 147, 204, 205
array, 37
array names, 37
ASIC *See* Application-specific integrated circuit (ASIC)
assertion, 20, 22
asynchronous, 27, 57, 71, 79, 214
    clocks, 148, 216

B
Best case, 194, 195
bidirects, 129
block, 13, 14
borrow, 103
Bottom-Up, 178, 181
break, 40
budgeting, 181

C
capture edge, 147, 148, 150, 151
capture flop, 147, 160
catch, 41
CDC *See* clock domain crossing (CDC)
cell_check, 196
cell_delay, 196
clock, 47, 52, 81, 92, 102, 103, 107, 124, 125, 134, 136, 159, 179, 197, 209
    divider, 57, 213
    domain, 71
    gating, 58
    multiplier, 58
    network, 21, 33, 88, 215
    path, 29, 197
    switchover, 50
    tree, 81, 125
clock domain crossing (CDC), 71, 75, 140, 152, 154
clock_fall, 102, 107
clock latency, 87
Clock network pessimism reduction, 33
Clock tree pessimism reduction, 33
clock tree synthesis (CTS), 3, 20, 81–83, 86, 88–90, 125
close, 41
combinational, 67, 78
    false path, 137
    paths, 157, 162
comment, 50, 53, 151
Commercial, 194
compiled, 36

configuration register, 139, 170
constraints, 7, 10, 19
continue, 40
Control flow, 37, 38
controllable, 2
coordinates, 206
corners, 183
CPF, 207
create_clock, 50, 69, 87, 160, 170, 210
create_generated_clock, 59, 87, 92, 152, 210, 211
create_pblock, 217
create_voltage_area, 206
critical path, 178
crosstalk, 76–77, 79
CTS *See* clock tree synthesis (CTS)
current, 198
current_design, 45, 200, 201
current_instance, 200–201, 204
cycles, 147

D
data, 197
    path, 29, 154, 197
datapath_only, 216
delay, 193, 197
    calculation, 26
derating, 122, 196
    factor, 196
derived clock, 57, 59
design for testability (DFT), 2
Design rule checking (DRC), 4
DFT *See* design for testability (DFT)
directive, 20, 21
divide_by, 62, 68, 211
dont care, 204
DRC *See* Design rule checking (DRC)
drive, 12, 23, 119, 122
driver, 50, 122
drive-strength, 118
-duty_cycle, 48, 64
dynamically sensitized false path, 139
dynamic power, 207

E
early, 89, 197
early analysis, 30
early late analysis, 30
early path, 88
edges, 61, 62, 65
edge_shift, 65

else, 40
elseif, 40
end, 150
end clock, 148
end point, 26, 28, 134, 135, 153, 165
envelope constraints, 184
Equivalence checking, 3
escape character, 198
exit, 41
expr, 38
Expressions, 38
extraction, 128

F
fall, 82, 85, 88, 103, 107, 118, 120, 125, 135, 146, 197
fall_from, 85, 135, 146
fall_through, 135, 146
fall_to, 85, 135, 146
fall triggered_high_pulse, 92
fall triggered_low_pulse, 92
false path, 131–132
Fast, 195
feedthrough, 162, 164, 178
field-programmable gate arrays (FPGA), 1, 4–6, 209
    prototyping, 6, 22
filters, 201
floorplan, 3
for, 40
foreach, 39, 40
Formal verification, 3
FPGA *See* field-programmable gate arrays (FPGA)
frequency, 48, 62, 210
from, 85, 132, 135, 143, 146
from_pin, 122
FSM, 151
full-timing gate-level simulation (FTGS), 6
functional mode, 168, 169
functional path, 28, 168

G
gated clock, 58
GDSII, 4
generated clock, 57, 78, 179, 187, 210, 213
get_cells, 45, 212
get_clocks, 45, 212
get_lib_cells, 45
get_lib_pins, 45
get_libs, 45

Index

get_nets, 45, 212
get_pins, 45, 212
get_ports, 45, 212
gets, 41
Gigabit Transceiver, 209
glitch, 139
group, 78
guard_band, 206

H
Handshake, 152
hardware description languages (HDL), 2, 9, 10, 19, 35
hierarchy delimiter, 198
hierarchy separator, 198
high pulse, 65
high transition, 48
hold, 29–30, 71, 84, 97, 101, 104, 107, 119, 121, 125, 127, 131, 136, 147, 148, 150, 158, 197
  edge, 147, 153
  multiplier, 147, 148, 150, 153
hold slack, 31
hsc, 199

I
IC, 1
ideal clock, 7, 81
ideal network, 93
if, 40
implementation, 10, 20
include_generated_clocks, 213, 214
incr, 41
Incremental, 110
Industrial, 194
info, 41
initialization, 139
input port, 96
input_transition_fall, 124
input_transition_rise, 124
insertion delay, 90
interclock, 83, 85, 86
interconnect, 95, 98, 104
interpretive, 36
intraclock, 83, 84, 86
invert, 62, 64
IP, 1, 164

J
jitter, 83–85

K
Karnaugh-map, 9

L
lappend, 38, 39
latch, 103, 107
late, 88, 197
late analysis, 30
latency, 92, 111
late path, 88
launch edge, 147–150, 153
launch flop, 147, 160
layout, 3
leakage power, 207
level_sensitive, 103, 107
level shifters, 206
lib_cell, 120
library, 19, 121, 197
List, 37, 38
load, 117, 193
LOC, 217
logical design, 1
logically_exclusive, 75, 78, 79, 186, 214
logical partition, 13
Logic block, 5
low pulse, 65
low transition, 48
lsearch, 38
LUT, 5
LVS, 4

M
master_clock, 60, 66, 211, 213
max, 82, 88, 97, 101, 103, 114, 119, 121, 125, 127, 131, 164, 195
max analysis, 30
memory, 155
metastability, 75
Military, 194
min, 82, 88, 97, 101, 103, 114, 119, 121, 125, 127, 131, 162, 195
min analysis, 30
min max analysis, 30
Mixed-mode clock manager (MMCM), 209, 215
mode, 167, 169
  analysis, 92, 172
Mode Merge, 173, 184
MTBF, 165
Multi cycle path, 131, 145
multi mode, 155

Multi-mode-multi-corner, 184
multiply_by, 62, 122, 211
multi-voltage, 205
Mutually exclusive clocks, 185

N
name, 50, 51, 53, 61, 67, 78
negative, 92, 153
negative delay, 114
negative edge, 102, 135
negative phase, 48
negative slack, 31
negative unate, 90
net_delay, 197
network latency, 87, 89, 105, 215
network_latency_included, 105, 108
no_design_rule, 123
non unate, 90
no_propagate, 93
n-transistor, 118, 193

O
observable, 2
On chip variation, 32, 195, 197
open, 41
operating conditions, 193, 213
operational modes, 183
Operators, 38
output port, 98
over pessimism, 33

P
parasitics, 7
Partially exclusive clocks, 186
partition, 14
Path breaking, 165
pblock, 217
PCIe, 209
performance, 35, 205
period, 47–48, 50, 51, 61, 62, 109, 160, 210
pessimism, 109
phase-locked loops (PLL), 58, 209, 210, 215
Physical design, 3
physically_exclusive, 76, 77, 79, 214
pin, 121
pin_load, 128
PLL *See* phase-locked loops (PLL)
positive, 92
positive edge, 102, 135
positive phase, 48
positive slack, 31

positive unate, 90
post-layout, 83, 86, 88
power, 23, 58, 147, 197, 205, 207
pre-layout, 83, 86, 88, 201
Procedures, 37, 40–41
Process, 193
procs, 40
p-transistor, 117, 193
Pull-Down, 121
Pull-Up, 121
pulse, 92
pulse width, 30, 49, 92
puts, 36, 37

R
recovery, 30, 136
register, 27, 157
removal, 30, 136
reset, 154
resistance, 117–119
resize_pblock, 217
return, 40
ripple counter, 57
rise, 82, 85, 88, 103, 107, 118, 120, 124, 125, 135, 145, 196
rise_from, 85, 135, 145
rise_through, 135, 145
rise_to, 85, 135, 145
rise_triggered_high_pulse, 92
rise_triggered_low_pulse, 92
routing, 23
RTL, 2, 3, 19, 209
RTL2GDSII, 4

S
scan, 2, 168, 179
  chain, 2
  mode, 168, 170
  path, 168
  shift, 168
SDF *See* Standard delay format (SDF)
sense, 90
sequential false path, 138
set, 37
set_case_analysis, 169, 173, 185, 204
set_clock_groups, 36, 75, 78, 140, 154, 165, 213
set_clock_latency, 88, 215
set_clock_sense, 90
set_clock_transition, 82, 125
set_clock_uncertainty, 84, 215
set_disable_timing, 143

set_drive, 117–119, 125, 129
set_driving_cell, 117, 120, 124, 129, 130
set_false_path, 131, 143, 157, 165, 179
set_fanout_load, 117, 127, 129
set_hierarchy_separator, 198–199
set_ideal_latency, 94
set_ideal_network, 93
set_ideal_transition, 94
set_input_delay, 54, 101, 106, 109, 111, 114, 123, 158, 160–164, 189
set_input_jitter, 215
set_input_transition, 117, 125, 130
set_level_shifter_strategy, 206
set_level_shifter_threshold, 206
set_level_threshold, 206
set_load, 117, 122, 127, 129, 130, 198
set_logic_dc, 204
set_logic_one, 204
set_logic_zero, 204
set_max_area, 204
set_max_delay, 157, 161–163, 165, 216
set_max_dynamic_power, 207
set_max_leakage_power, 207
set_min_delay, 158, 161, 162, 165
set_multicycle_path, 132, 165
set_operating_conditions, 195
set_output_delay, 54, 101, 109, 114, 158, 160–164
set_port_fanout_number, 117, 126, 127
set_propagated_clock, 90, 215
set_property, 217
set_resistance, 198
set_system_jitter, 215
set_timing_derate, 196
set_units, 197
setup, 29, 30, 71, 84, 98, 103, 110, 119, 121, 125, 131, 136, 147, 148, 150, 197
    edge, 147, 153
    multiplier, 150, 153
    slack, 31
set_voltage, 196, 198, 205
set_wire_load_min_block_size, 202
set_wire_load_mode, 202
set_wire_load_model, 201
set_wire_load_selection_group, 203
signal integrity, 77, 213
simulation, 2, 22
single cycle, 145, 155
skew, 3, 7, 83, 84, 86, 114, 216
slack, 31, 33, 110, 112, 181, 216
slew, 81, 82, 116, 193
    degradation, 126
Slow, 195

SoC, 57
source, 41, 60, 66, 88, 89, 215
    clock, 149
    latency, 87, 105
    synchronous, 67, 152
source_latency_included, 105, 108
source_object, 60
split, 38
STA See Static timing analysis (STA)
Standard delay format (SDF), 26
standard load, 127, 129
start, 149, 150
start clock, 148, 150
start point, 26, 28, 134, 153, 165
statement, 19, 21
Static timing analysis (STA), 16
stop_propagation, 92
strengths, 193
Strobe, 152
-subtract_pin_load, 128
synchronizer, 75, 152, 164
synchronous, 47
    clocks, 57, 71, 79, 83, 150, 213
Synopsys design constraints (SDC), 3
synthesis, 2, 9, 20
system synchronous, 152
SystemVerilog, 2

T
tapeout, 4
Tcl See Tool Command Language (Tcl)
Temperature, 193
threshold, 81, 82
through, 132, 133, 135, 145
Timing Analysis, 110
timing closure, 24, 177
    iteration, 173
Timing Exceptions, 22, 131
timing path, 26
timing report, 110
to, 85, 132, 133, 135, 143, 145
Tool Command Language (Tcl), 35
Top-Down, 177, 181
Tracks, 5
transition, 19, 81, 116, 124, 197
Typical case, 194

U
unate, 90
Uncertainty, 83, 110, 112, 215
uniquification, 181

Units, 197–198
UPF, 207

V
variables, 37
Verilog, 2, 198
VHDL, 2
victim, 76, 79
virtual clock, 54, 102, 107, 142, 159–160
Voltage, 193
voltage island, 205–206

W
waveform, 50, 51, 61
WEIGHT, 217
while, 40
wire_load, 128
Wire load models, 201
Worst case, 194, 195

X
XDC, 209
Xilinx, 209